There is no way

Seven Gates in an effortless way . . .

Going there is perilous . . . Amidst the silence, solitude, and darkness of the preliminary Gates, faces rise up to greet us from we know not where, to threaten us with unimaginable terrors.

There will be pain and the gnashing of teeth. The shamanic quest can promise nothing less than that, for if it did it would reward the initiate with nothing more than a hearty handshake and a funny hat, like the lodge brothers of so many '50s sitcoms.

Those of us who are drawn to the Quest are those of us who are propelled by "personal demons," and they must be met and neutralized.

For, after the personal demons have been mastered, other Demons await: The Gods of the Other Side.

The Ancient Ones.

Books Written or Edited by
Simon

THE GATES OF THE NECRONOMICON

SIMON

AVON BOOKS
An Imprint of HarperCollins*Publishers*

AVON BOOKS
An Imprint of HarperCollins *Publishers*
10 East 53rd Street
New York, New York 10022-5299

ISBN 978-0-06-089006-3
www.avonbooks.com

First Avon Books paperback printing: November 2006

Printed in the U.S.A.

HB 10.30.2023

To
Herman Slater

Father
of the Magickal Childe

Friend
of the Magickal Children

this work is gratefully dedicated

Contents

THE GATES OF THE NECRONOMICON

Foreword

This book is intended as a companion to, and explanation of, the *Necronomicon*. The *Necronomicon* is a work of occultism, translated from the Greek which was itself a probable translation of an Arabic language original. The reader should be advised that the *Necronomicon* is the center of heated controversy: there are many who assume or insist that the *Necronomicon* is a hoax, a mish-mash of ancient religious and occult practice packaged under the name of a book that was invented by the father of Gothic horror, H.P. Lovecraft.

There are others who insist, just as strongly, that the book is a powerful occult manual.

Often, the critics of the book are also those who claim that the *Necronomicon* is powerful, regardless of whether or not it is a hoax.

Originally published in 1977, it is arguably the best-selling grimoire—occult manual—in the world today, and this has caused even more consternation among the book's critics. Those who are interested in the lineaments of this controversy can consult my *Dead Names: The Dark History of the Necronomicon* for more detail.

The thesis of this book can be summed up in a few lines, as follows:

At some remote time and place in antiquity, an event occurred of such importance that it has entered the myth streams of various civilizations and cultures. This event involved the sudden appearance of a superhuman being or beings who taught tribes of illiterate peoples the basic requirements of any civilization: writing, astronomy, agriculture, architecture. Also a part of this exoteric teaching was an esoteric teaching, which described a nexus of death, the stars, the "astral" body, and what we now call the supernatural. Astronomy and architecture among the ancient peoples were intimately involved in this higher—or simply Other—purpose: the alignment of tombs and palaces to certain astronomical phenomena. The phenomenon most often referred to in these testaments of stone was a Celestial Gate. Although this Gate was entered at the time of death and lead the deceased to another World, it could also be opened by the living. But the Gate has—or had—an analogue: an anti-Gate, perhaps, through which malevolent forces could enter the World from the other side. These two Gates may have been one and the same, and improper use of the Gate might have lead to disastrous consequences.

This thesis is basic, not only to this book, but to virtually every occult or spiritual organization anywhere, anytime, on Earth. Its residues can be found in places as diverse as Aztec Mexico, Daoist China, ancient Egypt, and the lost civilization of Sumer . . . not to mention in the rituals of the Golden Dawn, of medieval European ceremonial magick, the Tarot, and the Enochian system received by Dr. John Dee. It is reflected, also, in the writings of Aleister Crowley as well as in the *Book of the Law*. It is also, of course, the cornerstone of the *Necronomicon*.

Using the above thesis as a starting point, and providing many examples from literature both sacred and secular, the author will further propose that the Gate spoken of by the ancients is an actual "place": that is, it exists, either as an event, or as a physical location, or—more probably—as a combination of the two.

And, the author will take this thesis one step further and provide tables and charts that will suggest the "location" of this Gate, along with instructions as to how it might be opened.

Naturally, this information is complicated and detailed. The background information required for an understanding of this important discovery is of necessity based on a knowledge of, or familiarity with, many languages (some long dead), mythologies, and sciences. It is the author's hope that he has made this information as clear as possible. A comprehensive bibliography is included so that interested or critical readers may follow the author's line of reasoning for themselves.

The existence of a Gate or Gates in the *Necronomicon* is central to its entire mythos. The Mad Arab speaks of the Gates several times in his opening remarks, and then goes on to give quite emphatic instructions as to how the Gate should be opened, and when, and accompanied by what words, diagrams, etc. He also insists that, once having begun the process of going through the Gates, one should not stop along the way but continue straight through until all seven have been passed. The reason is clear: passing through only a few of the Gates is enough to let something Other *in*. Should a malicious entity of some sort manage to sneak past the operator and into the World, the operator's lack of experience on the higher planes (represented by the remainder of the Gates) would be fatal: not only to him- or herself, but to society in general.

The Gates are *related* to the seven "philosophical" planets, that is: Mercury, Venus, Mars, Jupiter, and Saturn, together with the two luminaries, the Sun and the Moon. As we all know, the Sun and Moon are not planets. The Sun is a star, and the Moon is the earth's satellite. In the old days, it was believed that the heavens revolved around the earth. This geocentric view of the cosmos gave rise to astrology, which became modern astronomy when it lost its humanist characteristics and placed the Sun at the center of the solar system, and the earth in revolution around it along with the other planets. This switch from the earth being the center of existence to the Sun triggered a grave disturbance in the human psyche, one from which we have yet to recover. It goes hand-in-hand with Darwinian evolution, and the resulting birth of a new science of artificial intelligence that implies that there is no thinking that a human can do that a machine could not, one day, do better. As thinking (i.e., consciousness) is—to our eyes—all that really sets us apart from the animals, we are in danger of not being in the center of even our *own* little universes anymore. We, as humans, become relegated to accidents of nature. And, as such, we lose God.

In losing God, we lose heaven and hell. We lose the afterlife. We lose reincarnation.

But what if the seven stages of awareness, the seven steps up the ladder of lights—the Seven Gates—do not pertain to the seven "planets" at all? What if the ancients knew all along that there was a higher order in the universe, some realm of wisdom and power far beyond our small solar system in a corner of a galaxy in a corner of the macrocosm? What if the seven "planets" were a blind, something to occupy the activity of the astrologers and the other "fatalists" who believed that humanity's destiny was written in the stars and could not be changed? A determinist perspective designed to ensure the people's docile acceptance of the circumstances of their existence: the king, the state, the land?

What if the *Necronomicon* itself hints darkly at another science, another "astrology," if you will, running parallel with the astrology we already know and yet never really touching it: not operationally, not ideologically? What if, in this vast mechanism of time, space, and destiny there was a "ghost in the machine"? A loophole? A way Out?

Gates of the Necronomicon explores that possibility as it analyzes the initiatory procedure of the *Necronomicon*, with special reference both to its "calendar" and to the method of psychic self-structuring it details. It points us back to a War that took place—on earth? or in the heavens?—in some distant past and how battle lines were drawn in the earliest recorded civilizations of our planet. A War between two Stars, whose battle is memorialized even today among the sects and cults that proliferate on our World. The weapons of that conflict are the tools of the magician and the alchemist, the sage and the soothsayer, where immortality is only the first step in choosing sides.

For our study, we will find ourselves delving into the mysteries of Sumer, Egypt, India, Tibet, China, and Mexico. We will examine closely the literature of magick, alchemy, astrology, and yoga. Our goal will be to identify the Gate, and the means for opening it, and the reasons why we should.

Included as the introductory essay to this book is something I call a "Prolegomena to the Study of Occultism." It was designed to answer many of the questions I usually get during my classes and seminars, and in the letters I receive. Those of you who are already quite comfortable in the world of occultism—especially of the ceremonial magick variety—need not concern yourselves with its contents and can proceed directly to Chapter 0. If you are curious, however, as to how magick, reality, and sanity may be related and how secret societies can serve to protect the individual magician from harm in the World, then this Prolegomena may hold some interest for you. It is both an Introduction to Magick, and the Manifesto of one Magician who has traveled the planes of existence and come back with a message . . . and a warning.

THE GATES OF THE NECRONOMICON

Prolegomena to a Study of Occultism

Reality, Fantasy, and Sanity

One of the issues that plagues any thinking person who becomes involved—however innocently, however briefly—in the study of psychic phenomena, occultism, ceremonial magic, astrology, or any of the other "pseudosciences" that get lumped in together with the above, is the challenge of *reality.* "Is it *really* happening, or is it just in my mind?" is one of the first questions any teacher of the occult must answer. "It isn't *real*, it's all in your mind," is one of the first snorts of derision that anyone involved in occultism must suffer. "It's nothing but fantasy," they say. They know this because they have been *told* that it is fantasy, that it is *not* real. They have been told this by their teachers, by pop scientists like Sagan on television, or by their friends. They themselves have not investigated the paranormal at all. Quite often, they are not equipped for such an investigation. What they "know" is what they have been told. You will find that the dullest, most functionally illiterate mental mushroom has a very definite, very "scientific" view on one thing: the impossibility of any kind of psychic phenomena. "There ain't no such things as ghosts," might typify one of these brilliant scientific assessments of centuries of human experience. And anyone who "believes" in ghosts is crazy. By linking the twin concepts of *belief* and the *paranormal* we arrive at a cogent example of the use of language to alter perception, for we either "believe" or don't "believe" in ghosts, magic, God, the Devil. There is simply no corresponding approach to plane geometry, the square root of minus 1, or the genetic code. One never asks if one "believes" in the Pythagorean theorem or in any Euclidean theorem (or, for that matter, no one is ever asked whether or not they "believe" that the circumference of a circle contains exactly 360 degrees, even though that number comes down to us from Babylonian mythology and is not the result of "scientific observation").

Rather than waste a great deal of time in attempting, within these pages, a general defense of an occult "belief system," the author would like to point out just a single important element in this debate that generally goes unremarked: the origin of the very words "reality" and "fantasy" and what they actually imply when used in this, or any, discussion.

Reality comes from the Indo-European root *reg*, from which is derived a host of words meaning "straight, direct, in a straight line, ruler (in the sense of a measuring device)" and "king, kingdom, ruler (in the sense of a political leader)"; words such as *regal, royal, regulate, realm*, and *rank. Reality*, therefore, is intimately linked to the idea of the king. Those readers familiar with ancient history will remember that the king—or the pharaoh—is the person responsible for setting the limits of the sacred precincts, for the design and orientation of the city-state. The king decided what was *real*. Anything outside that area was not *real*. Was not *of the king*. Was not subject to measurement, to the *ruler*. Therefore, when people ask you, "What does it *really* mean?" what they are asking is, "What does *the king* say it is?" When they ask, "Is it *real*?" they are asking, "Is it part of the kingdom? Part of the *realm*?"

Fantasy comes from the Indo-European root *bha*, from which is derived another long list of words meaning "to shine, to be brought into the light." The implication, of course, is that the subject being brought into the light had existed previously in darkness, in the *realm* outside that of the king: the part that was *not* subject to the measurement of the ruler.

But to "bring something to light" is not to question its existence, but merely to decide whether or not something should be permitted within the realm of the real, the realm of the ruler (of that which is measurable), or to leave it outside the realm, in the "outer darkness."

Hence, the question of reality—of what is, and is not, real—is properly one of politics. The idea of reality is just as much a part of the *civil* law as the idea of "insanity," which is also a civil, and not a scientific, term. Reality, fantasy, and insanity are determined by the king (in a monarchy), by the state (in a fascist, communist, or other totalitarian society), or by the general consensus (in some other forms of government, or in your own neighborhood). Thus were the parameters of "reality"—of the "real estate"—set down at the four corners of the realm a long time ago, and have never been questioned since then by the majority of the population. We are told that reality is determined—first of all—by the five ordinary senses. Objects or events that affect sight, sound, taste, touch, and smell can be judged "real" objects: i.e., they fall within the measurable limits of the realm. However, the testimony of only one person as to whether such an object or event did affect any of these senses is not sufficient to declare it *real* (unless that person is the king). What is required is a general consensus of agreement on the part of the *rulers* that the senses were affected in the way described by the object or event thus identified. What happens when someone sees a ghost, however? Or hears voices? Or smells the sulfurous fumes of hell? Or feels the brush of an ethereal hand? Or bites down on the sacrificial bread and tastes flesh? That is when the experience of the individual splits off from the experience of the king or of the group: when it ceases to be *reality*. In this way, even our dreams have become *unreal* even though there can be absolutely no doubt that they exist.

In a totalitarian government—as was the Soviet Union until only recently—the state decides who is sane and who is insane, and commits many political prisoners to asylums for the mentally ill. In other words, the prisoners have questioned the view—or the *parameters*—of reality as perceived by the state and are, therefore, insane. They cannot tell "right" from "wrong," reality from fantasy. The Soviets realized very early on the connection that exists between concepts such as "sanity" and "the state." The consensus, therefore, was not a consensus of the general population but of the ruling class in a Soviet-style society: the Politburo, perhaps, or the Central Committee, which would delegate responsibility for determining sanity and insanity to an agency of the government, such as the KGB, for instance.

In the United States we may cite a similar example. In nearly every case where an assassin has tried to kill a President or a candidate for the Presidency, he has been deemed "insane." It is always a "*crazed, lone* assassin"—an Oswald or a Sirhan in the case of the Kennedys—who has attempted the deed. The message is: to attack the Presidency can only be the act of a madman, a loner with no social ties, someone outside the *realm*, someone not *sane*.

Manson was insane; Lieutenant Calley was not. Calley was part of the government, *ergo*, his act (though admittedly wrong) was not the act of an insane person. Manson was an ex-convict, a member of that society which is *ipso facto* outside the *realm*. His acts were those of a crazy man. His followers were crazy. Kooks. Psychos. Brainwashed. Cultoids. For does it not say in the Old Testament, first Book of Samuel, Chapter 15, Verse 23: "For rebellion is as the sin of witchcraft"? The fact that Manson's most famous victim, Sharon Tate, was the daughter of an Army colonel and intelligence officer in Vietnam only underlines the symmetry we are describing.

Lieutenant Calley—also following orders—was quite

sane. After all, he was engaged in bringing the realm into the lands of darkness, of extending the rulership of his President into the jungle and thus defeating the fantasy of the Vietnamese.

Manson, on the other hand, was bringing his fantasy into the realm. Although Manson is guilty of murder, as are those of his followers who also partook of the slaughters of that August week, the question we are looking at now is not that of murder, for Calley also murdered. And Noriega murdered. And General Pinochet. And President Stroessner. And Ferdinand Marcos. And the Duvaliers, *pere et fils*.

What we are questioning here is the popular use and understanding of the word *sanity*.

Sane comes from an Italic root meaning "healthy." We find it in such words as *sanitary*. *Sane* does not mean, in its pure sense, the ability to determine what is real and what is fantasy, nor even what is right and what is wrong, what is good and what is evil. Although the courts interpret *sanity* as the ability to tell "right" from "wrong," what it *does* mean is "healthy." The implication is that those in agreement with the king or the government or the consensus are healthy, are clean. Those not in agreement are unhealthy, sick, unclean. *Insane*. Mentally *ill*.

(The moral dimension of health and sickness has remained one of the legacies of medieval theology and civics hardest to extirpate from the modern civil code, and has led to excesses in the courts such that the American legal system itself is under attack from parties in all colors of the political spectrum. The sudden increase in the number of serial killers in the United States—murderers who are, almost by definition, "insane"—has provided the legal system with a Gordian knot of complex, interrelated legal, medical, scientific, philosophical, and moral issues that it is not equipped to handle.)

Therefore, those who take the occult and the paranormal seriously are committing what is essentially a political act, and are in danger of being considered *insane*, at the very worst, or at best being simply out of touch with *reality*. One lives in a *fantasy* world, bringing to light that which the king has determined should be left in darkness, left outside the *realm*.

You ask whether what you experience *really* happens, or only happens "inside your head," in your mind. The answer is this: until society in general, or the king in particular, determines that what you experience—inside your head, or out, if there are such places—falls within the measurable *realm*, what you experience is *fantasy*. Politically speaking, acts of occultism are political acts subversive of *reality* for they *bring to light* that which the king, the ruler, has determined should be left in darkness. Therefore, the occultist tampers with his or her own *sanity*: that is, he or she is in danger of being considered incapable of discriminating between reality and fantasy, of knowing "right" from "wrong" in either the moral sense or the scientific sense. The occultist is either evil or insane. There are no other options. For *rebellion* is as the *sin* of *witchcraft*.

Therefore, until the king, or the government, or the Politburo, or the State, tells us that what happens inside our heads is *real*, we are not dealing with *reality*. But that's okay with me. Once it becomes part of *reality*, part of the *realm*, the king will find some way to *regulate* it, maybe even to tax it. Therefore, I am proud to declare to one and all that what I do is *not* real, is pure fantasy, and that yes, I *am* evil . . . and, like the Mad Arab, very probably insane.

The Secret Society

Was there a moment in human history that could be called the "dawning of consciousness"? If so, then was it a move from an unconscious state to a conscious one? From darkness into light? A moment of, shall we say, fantasy?

If so, then we may posit that other members of the mammal family who have not yet evolved to the point of being human may be living in an unconscious state similar to the state that exists in our own dreams, or a preconscious state in which the dream state and the state of perceived, or consensus, reality are mixed. (That is, a condition preexisting the "State" or a ruler who could measure the limits of the realm.) Hence, the witch's familiar: an animal that could detect unseen influences more easily than the witch and possibly protect the witch against them. A preconscious being unburdened by the requirements of a State.

Does consciousness depend upon language? If so, then the efforts of some scientists to teach apes and dolphins to speak would be tantamount to giving them consciousness. As they live—half in our world through the use of speech and half in their own world through the use of their own instincts and behavioral patterns—they become split into two sectors: a conscious one and an unconscious one, just as we are. We can see that, for an animal, consciousness as we perceive it is the result of interacting with human society.

Then, we can understand the ancient prerequisites for reaching the unconscious state through isolation, celibacy, prolonged periods of concentration, nocturnal rituals and meditation: all activities and circumstances designed to put as much distance between society (the State, the Realm) and the individual as possible.

Thus, the goal of every secret society should be the preparation of the individual for this *solitary* quest; i.e., it may be impossible for a group of people to carry out the ultimate journey into the unconscious together, at the same time, by definition—for a group is a small society, and society implies consciousness.

During the journey, the secret society may provide security for the initiate—by keeping a careful watch over the body, by reviewing (as much as possible) the data as it is collected during the journey, by protecting the body from distraction and from the scrutiny of the general society.

And after the journey the secret society plays an important role in the debriefing of the individual, the validation of his or her journey, and the gradual "decompression" of that individual so that he or she may rejoin the ranks of the greater society in relative safety.

In this manner, the secret society is a way station both out of and into general society for the individual on a spiritual quest. Vows of secrecy should be strictly maintained: their importance cannot be overemphasized. The integrity of the secret society rests in the society's ability to guarantee the safety of its individual members. Typically, in every general society, those who attempt to "go inward" or approach the unconscious risk the disapproval of, and persecution by, their neighbors, as the Quest is an antisocial act. It is an act taken on independently of society, an act not sanctioned by the ruler, a communication outside the realm, a journey beyond the royal purview. It is therefore an act rife with political overtones. Those who betray the trust of the members of a secret society and expose the acts of its members to the outside world work for the general society, for the realm, and can be considered spies or saboteurs for the king. The exception to this rule is when the acts are exposed in such a way as to be incomprehensible to any but other initiates, in which case this dissemination of information can be quite properly understood as a kind of ironic "disinformation" and serves to further protect the secret society from the scrutiny of the general society, the realm. At certain levels or stages of initiation the data is extralingual anyway, beyond verbal description, and cannot be betrayed or revealed no matter how intense the desire.

Basically, the data received from such journeys may be published, discussed, or otherwise disseminated if that is the consent of the secret society concerned; however, the identities of the individual members who have undertaken these "subversive" acts should be protected at all costs. Hence, the need for aliases and "Order names": names taken by the individual members reflective of their "secret" identities and/or goals. Members who seek publicity among the general society for their deeds or their position within the secret society generally wish the approval and recognition of the realm; the sincerity of these members is suspect, and their desire is foolish because the realm has condemned their activities and will never offer its approval for acts it considers subversive of "reality." They should be shunned, because they may easily expose the identities and deeds of other members of the society, either to ingratiate themselves with the realm or to take revenge on the secret society for imagined hurts. Neither one thing nor another, they occupy much the same position between these two forms of society that informers enjoy between the criminal world and the legal authorities.

Microcosm and Macrocosm, Reality and Fantasy, and the Gate Between: the Seventeen Axioms

As occultists, we need an understanding of the environment in which we operate before we can accomplish what we set out to do. Usually, this type of environment—what I am calling here a "cosmology"—is generally glossed over in the literature or else so minutely described by one or another of the sects or secret societies—and thus so dependent on its own, peculiar cultural formulas—that it might as well be omitted entirely (unless you are a communicant of one of those sects or secret societies).

This cosmology is based on some very simple observations of the contours of the macrocosm/microcosm in the literature of various religious and mystical authorities from all races, all cultures. Basically, it can be set out like this:

I. There is a world we see, and a world we don't see.

II. Actions that take place in the world we see can influence circumstances in the world we don't see. Magic is an example of this.

III. Actions that take place in the world we don't see influence circumstances in the world we see. Astrology and some religions are examples of this.

IV. The world we see is a visible emblem of the world we don't see. Individual phenomena in the world we see can be related directly to phenomena in the world we don't see. Everything in the world we see has its counterpart in the world we don't see. This is the doctrine of correspondences.

V. The human life cycle begins at birth, lasts through life, and ends at death. At death, another phase of existence *may be experienced.* This phase of existence is referred to in most cultures as the *Underworld.*

VI. Since our bodies decay or are destroyed by fire or water or some other form of burial, if there is another phase of existence beyond Death, it must be experienced by a form of our bodies that we can't see.

VII. If there is a form of our bodies that we can't see, it belongs to the world we don't see. As such, its existence in the world we don't see must be counterminous with the existence of the bodies we see in the world we see.

VIII. Since we cannot see the Underworld, the Underworld belongs to the realm of the world we don't see. (See Axioms II and III above.) We can influence this Underworld, and it can influence us. Therefore, a medium for influencing the Underworld may be the invisible form of our bodies. It may also be the medium the Underworld uses to influence us. This Underworld may also be related to the unseen organs of our consciousness, what have been called the unconscious, the subconscious, id, anima, shadow, etc. According to the psychologists, what occurs in the unconscious mind influences events in the conscious mind, and vice versa, which is a modern way of saying the same thing.

IX. The world we see is related to Light, which makes things visible, measurable. The world we don't see is

related to Darkness, which makes the world we see invisible and therefore unmeasurable. The time of the world we see is Day. The time of the world we don't see is Night. The time we use for working—for serving the realm—is Day. The time we use for sleeping is Night. Therefore, **to visit the world we don't see, we must work at Night when the world we don't see becomes manifest.** This is also what psychology teaches when it speaks of the importance of dreams, which is a phenomenon experienced during sleep.

X. At Night, the stars and planets are visible as points of light. These forms of existence during the time proper to the world we don't see represent the phenomena we don't see that take place, among other places, in the Underworld. This is according to the doctrine of correspondences, mentioned above. **They can represent the bodies we don't see: what some cultures call the "astral body" or the "star self"** in a reference to its relation to the stars. They are significators of our unseen selves and of other unseen phenomena. They never die, but revolve eternally in the Underworld. This *implies* that, if we have unseen bodies or selves, they are also eternal.

XI. As these lights are actually themselves seen, then they must have unseen counterparts in the world we can't see. Therefore, there are three types of existence that concern us in this place: the world we can't see, the world we can see, and the world between them, which is represented by the stars of the nighttime sky—what has been called the *"astral plane."*

XII. For the above to be useful, **there must be points of tangence between the visible world and the invisible world: a channel of communication** from one "world" to the other. **This channel is what is referred to as a "gate."**

XIII. There are functions within our own bodies that remain invisible to us. This has given rise to the theory of invisible centers within the visible body that control various phenomena. These may, or may not, be counterminous with postulated invisible centers in the "astral" body. **These may, or may not, be reached via the same type of "gate"** mentioned in XII above.

XIV. The gate used to pass from the visible world to the invisible world outside our bodies may be the same gate as that used to travel from the visible centers to the invisible centers within our bodies.

XV. Whatever the case, **if there is no visible sign of a phenomenon, then that phenomenon is deemed not to have an invisible existence,** either. In other words, for every invisible thing that exists there must be a visible thing. This may not, of course, be true. But it is the safest theory to accept for now. A corollary is: to create an invisible phenomenon one must first create its visible counterpart. This has been known as idolatry. Or, to create a visible phenomenon, one must first create its invisible counterpart. This has been known as casting a spell. If an attempt is made to create an invisible phenomenon, but no visible phenomenon results, then that attempt is deemed a failure, i.e., the spell did not work. But if a visible thing is created, then its invisible counterpart is simultaneously created; i.e., the god of whom we have made a statue actually "exists." Whether or not the god is powerful is determined by the power of the creator(s) of that god.

XVI. Communication between the two "worlds" can be weak or strong. A weak communication is what is known as a dream. A strong communication is that effected by deliberate acts of the conscious mind on the unconscious (magic, spells, etc.) or by manipulation of the conscious mind by the unconscious (psychoses, hallucinations, visions, etc.). Both of these types of strong communication are politically suspect and can deliver their host or practitioner to prison or asylum.

XVII. Therefore, **it is advisable to practice strong communication through the medium of magic** because it restricts the operation of the channel or "gate" to control by the conscious mind, which can conceal its use from the realm and synchronize the opening of the "gate" in accordance to a calendar of safety, i.e., when the king is asleep.

Tantra, Sexuality, and the Black Mass

You can legislate reality, but you cannot legislate experience. You can legislate marriage, but you cannot legislate love.

Tantric love—as it is understood by the popular press—is an act of sexual union within the framework of a religious ritual designed to bring both parties to enlightenment or to any of the other goals mentioned above. Tantric love is illegal love. Ideally, it is performed with a woman married to someone else. It is directly in the same tradition as that of the troubadors of the Middle Ages, and, indeed, Tantra may have influenced the cult of the troubadour. Tantra turns organized religion—the legal, authorized form of worship that exists within the parameters of reality as determined by its rulers—on its ear. It incorporates the illegal into a new definition of the legal. It sanctifies the unholy and, what is more, becomes the ultimate subversive act because it violates the rules of "real estate" and, at the same time, is performed in secret and leaves no trace—unlike other forms of rebellion. Further, its goal is liberation from the realm, from the mundane world, through the influence of the senses themselves: the very senses the realm considers the means of ordering and determining reality.

* * *

The Christian world—under the Catholic Popes—became increasingly urban and "civilized" once their hold on political power in Rome was consolidated. The rural population—the people who lived in the country and who were far from the cultural and legal manipulations of the Roman popes—were referred to as "pagan," a word derived from the Latin *paganus*: the countryside. Hence, a division of the world into the Christian (and urban) and the Pagan (and rural). This division would soon take on tremendous political implications during the Inquisition, when millions of "pagans" were persecuted, jailed, and executed for either heresy (disagreement with the terms and parameters of reality) or for trafficking with the Devil (the Lord of the Underworld). Naturally, their "real estate" was also seized by the Church or by individual inquisitors as part of the persecutions. By putting themselves—or finding themselves—outside "reality," due either to their ignorance of the new terms of reality or to their willful allegiance to the terms of an alternate reality (their native religion), the Pagans lost all title to "real estate."

Similarly, the experience of "love" was generally frowned upon by the ecclesiastical authorities as it could not be seen, or measured, and hence incapable of being placed solidly within the realm (which is the world of the visible and measurable, the "world-we-see"). Further, unions based on "love" demonstrated allegiance to the terms of an "alternate reality," one opposed to the carefully constructed scheme of documents and approvals. Marriages were arranged between noblemen and their families, who wished to consolidate political power and influence . . . all down to the level of farmers with property they wished to merge. Marriage was an extension of *real estate*, as it was tied inextricably to the land and ownership of property (as it is today). When "love" became fashionable, it was properly the domain of the troubadors, who sang almost without exception of "illicit" love: of that between a married woman (usually a "lady" and therefore a member of the ruling class, of the ruling family of the realm) and a man not her husband. The concept of love being related to fantasy, and therefore outside the realm, has become part of the accepted view of life held by civilizations as different as modern Japan and medieval France. The idea of love being somehow illegal has never left Western civilization, although we rationalize our own marriages as matches made in heaven (once legitimized by the king, and not before) and thus similarly eternal. Human expectations are therefore raised to illogical limits, and the devastation that occurs when human experience teaches us otherwise is inescapable . . . and needless.

By accepting the court's approval of romantic liaisons in the legal act of marriage, the two parties involved tacitly accept the conditions of reality as embodied by the laws of the realm. This extends to definitions of love, which the law has tried desperately to control, as love is a force with the almost limitless ability to transform social relationships and power structures. (Anyone who doesn't agree has merely to read the newspapers. Everyone from Presidential candidates to TV evangelists has been destroyed because of "love": because of the inability to reconcile the legal definition of reality with individual experience.)

This attempt at equating the legislation of emotional and sexual pairings in the marriage contract to moral and spiritual "contracts" through creative defining of love and the arbitrary assignment of its operating parameters, is doomed to failure, as it has always been, since the earliest recorded instances of adultery, fornication, homosexuality, and all the other "forbidden" forms of love. These forms were forbidden, as they acknowledged the existence of a force that operated quite effectively outside the realm. *Pro forma* government approvals of homosexual, adulterous, and other liaisons would have been impossible to enforce, as there was no way to control real estate effectively through such unions. With a homosexual marriage, there would have been no legal offspring to represent the consolidation of two pieces of property: i.e., no "new entity" emblematic of the union of two previous entities. In an adulterous liaison, the question of property would arise once again, and of the legitimacy of any potential offspring. On and on, through all forms of "forbidden" liaison, we can see the quandary governments found themselves in when legislating sexual pairings.

Many governments found a way out of the conundrum by tacitly accepting the existence of concubines, prostitutes, etc., as quasilegal entities entitled only to a form of compensation for services rendered, and not legally entitled to real estate or title to other property, and whose offspring would not be "legitimate": that is, not in the line of succession to political or other temporal power in the realm. "Illegitimate" children existed outside the realm, in the fantasy world (were, in a sense, pagans), with no ties to property or "real estate" of any kind, and have been called "love children" in recognition of the fact that they represent this unlegislatable force that defies the king, the rulers, and the governments that dictate the parameters of reality. As the "love child" cannot fit within the rigid requirements of legal title and rules of succession, they are ostracized as the result of an "evil" act: an act which, by its very nature, questions the authority of the king, of the realm, and of the parameters of reality. In the megalithic culture of the Malekulans, for instance, illegitimate children were raised in separate communities to become human sacrifices once they reached puberty.

It is no wonder that Charles Manson is the fruit of just such a liaison, an illegitimate child who suffered immeasurably because of his status as a "nonhuman," and who remains as the symbol of all that is wrong with society. The realm turned on Charles Manson with a vengeance, when he was still a child, and he repaid the moral debt with interest. In a sense, he *is* the anti-Christ the media have made him out to be, for he is not the kind of victim society wants to see: the Christ victim who accepts his fate silently and who bears his cross in suffering, pain, and humiliation for the rest of *us*, and who waits patiently for *us* to accept *him* at our convenience. No; Manson represents the victim who is tired of being victimized, and who turns on society with fire and sword. Manson refuses to be the quiet, acquiescent victim. In this, he is as different from Christ and the Christian ethic as possible. He is a lot closer to St. Peter, who struck the Roman soldier with his sword and who was waiting for a military messiah to deliver the Jews from the Roman occupiers. Like St. Peter, we may think of Manson's symbol as less the swastika than it is the inverted cross of the Satanists on which St. Peter was crucified.

Tantra, illicit love, and the anti-Christ. All of these symbols become unified—albeit erratically and in a disorganized fashion most of the time—in the obscene parody of the Catholic liturgy known as the Black Mass. In this rite, worship of the Devil (Lord of the Underworld, of the Unseen World, of Fantasy) is combined with a reversal of Catholic values and laws. In the classic example, a nude woman (preferably a prostitute) is draped on top of the altar and serves as the surface on which the rite is performed. The priest (preferably a genuine, defrocked Catholic priest) performs the ceremony, which includes celebrating the Catholic Mass backwards, even so far as to recite the various prayers in reverse, walking backwards, etc. The woman on the altar, in some examples, performs sexually with the priest and/or members of the congregation.

The Black Mass is an act of intense anger towards the Church, of revulsion for its principles and rituals, which is somehow combined with a performance of the sexual act. This idea of sex existing in opposition to the Church is nothing less than the Tantric view. Unfortunately, in the case of the Black Mass, there is no more serious attempt at entering the unseen world than there would be in, for instance, getting drunk or stoned. The benefits are largely momentary and not of lasting value. No one who has ever performed a Black Mass or participated in one has ever gone on record as saying that it had in any way enlightened them or brought them closer to spiritual goals. In negating the offices of the Church, the Black Mass ends its usefulness at the very threshhold of illumination; it's a tease. It brings the participants to the very edge of entering the unseen world and leaves them there without a map, without a compass. It is also so culturally loaded as to be ineffective for the majority of the world's population not unduly influenced by Catholicism. In this post–Vatican II age when Masses have already taken virtually every form imaginable, its utility as a means of freeing the unconscious is questionable.

Tantra, however, by violating the rules that bind nearly *every* society—the rules concerning marriage and sex outside marriage—is far more useful, particularly as there exists within its structure a coherent framework of mental and physical preparation and a complete cosmological system with a definite goal, enabling the participants to have some idea of where they are going and how to get there. As a rite, it consciously violates the rules of reality, of the realm, and propels its participants into the fantasy world where the identities and relationships imposed upon them by the realm disappear. Tantra is denounced by organized religion and by the government. It is socially unacceptable. This is not only because of the violation of the sexual taboo, but is also due to its violation of various dietary taboos. It is a conscious, perfectly conceived and patiently executed, subversive act; and it takes place within a goal-oriented framework that lifts it above a mindless act of blasphemy and anger against the State (a kind of philosophical vandalism) to a deliberate act of political sabotage (that is, rather than vandalism, the creation of a new realm with a new king and queen, an Underworld realm where fantasy becomes reality for the participants involved; a philosophical "cell" composed of members committed to the Revolution).

Scientific Method and the Skeptics

Anyone carefully reading the previous pages will be able to predict with accuracy what the author will say concerning scientific methods and the current crop of professional skeptics, including Sagan, Kreskin, et. al., who like to debunk astrology, the paranormal, etc.

In the first place, the available literature published by and about the skeptical organizations reveals their one-sided approach to the subject matter: they seek out straw men to knock down. The author has yet to see a careful investigation of astrology, for instance, that does not rely on the daily newspaper versions that all professional astrologers are quick to admit are so hopelessly general and vague as to be quite worthless. I am reminded of the *Cosmos* episode in which Sagan, for instance, reads off the predictions for various sun signs in a newspaper and appropriately ridicules them. It seems that it is only with the paranormal that science can afford to abandon scientific method, and instead condemn these practices as futile on the basis that current scientific theories (in the case of astrology, those pertinent to the gravitational and magnetic influences of planets and luminaries on the earth) do not permit their inclusion with the other "sciences." This attitude would have been familiar to Copernicus and Galileo.

In the second place, however, this author is all in favor of not permitting science to have any say whatsoever in the practice of astrology, yoga, meditation, ceremonial magick, or any other paranormal or supernaturally oriented discipline. It would be a mistake to permit these practices to exist as disciplines within the realm. Science complains about the effects of ESP, PK, and other forms of individual, personal psychic ability and powers as being "unmeasurable." Naturally, the "measurable" is automatically part of the "realm," of the "ruler." We prefer to contain these practices within our own sphere of the Underworld. Concerning Science, we can be thankful for the vaccines that have saved hundreds of thousands of individual lives, but at the same time gaze in horror at the damage that has been done to the entire planet, damage that may be irreversible. The disservice that Science has done to humanity can be calculated by the degree to which it has made us rely on the measurable, the seen, the tangible. All that our Science can do for us now is to measure the full parameters of the disasters that will visit this planet in the years to come. It is powerless to help us avert them. Only we can do that.

Summary

Occultists properly operate within the world of fantasy, and not reality. However, they understand these terms in a different way from the general population. They understand that reality and fantasy, and their corollaries—sanity and insanity—are political concepts and not entirely relevant to the acts they perform except as cautionary signals from the culture within which they must operate.

The only social support they can expect is from the secret society, insofar as that society supports their efforts and is pledged to conceal their processes and identities from the prying eyes of the State and its many loyal subjects.

The process they are utilizing involves opening a channel of communication called a "gate" between the visible world and a hypothetical invisible world. The visible world is the measurable one of reality. The invisible world contains events and data that, when brought to the light of the visible, are functions of the act of fantasy. Members of the general population either "believe" or "disbelieve" in these events and data. Occultists observe the events and data and record them for future investigation and analysis and possible use. Belief is not a factor.

This is the bare bones of occultism. Any description beyond this one is culturally loaded and may, or may not, be valid for individual readers, practitioners, believers. The cultural contours are important and powerful insofar as they permit the occultist to open the gate safely and easily, and to analyze the communication in a meaningful way, i.e., to integrate the data into his or her own cosmology in such a manner that he or she attains greater knowledge (information) and power (the ability to use this information to exist—and flourish—both inside and outside the realm).

The end result of all this pertains to a goal within the unseen world for which there is no single term appropriate to every occultist, i.e., the concept is culturally loaded and has been referred to as "unity with God," "immortality," "illumination," "enlightenment," "changing base metal into gold," etc. There are other terms, and perhaps even other levels, other goals, beyond these. If so, they are properly the content of the unseen world. This does not imply that there is no "visible world" counterpart to a particular goal but only that it has, so far, gone unrecognized in the visible world; unrecognized by the king and by the king's greatest weapon: the king's language.

Listen to the voices that come in the night, the voices of madness, the voices from the Underworld. Put away your preprogrammed reactions of horror and derision for once, just this time, and listen to the message as well as the medium. Put down your guns, which you pick up in fear and hatred, and raise your wands high in the air, in joy and confidence. Follow no prophets, no messiahs. Do not trust them, for they will stand between you and God, between you and yourself.

God is the only safe thing to be.

CHAPTER 0: Necronomicon: The First Thirty Years

The tale of the discovery and publication of the *Necronomicon* has been detailed in *Dead Names: The Dark History of the Necronomicon*, but a summary might be in order.

The story has pretty much passed into legend by now.

Two monks of an Eastern Orthodox Church were arrested by Federal authorities in the early 1970s for the theft of rare books from private collections and university libraries around the United States and Canada. It was the largest rare book heist in the nation's history. Involved in the investigation was another monk—an Abbot, no less—who infiltrated the rare book ring and reported back to the Church leadership. Several months later the two monks were indicted. They have since served time in various federal institutions.

The rare books that were stolen numbered in the hundreds, if not the thousands. Some say the full extent of the damage will never be known. Many of the books were "cannibalized" for prints and maps, which were torn out and sold separately to collectors. In some cases, according to the story, a van was backed up to the residence or library in question and loads of books were piled in.

One of these books was the *Necronomicon*.

The Abbot who performed the initial, internal investigation was known in other circles as a ceremonial magician. In 1972 he began teaching classes in old-fashioned, medieval ceremonial magic in a hotel room in Brooklyn Heights, about five blocks from H.P. Lovecraft's old apartment. The following year he designed the Ceremonial Magic exhibit at the Museum of American Folk Art in Manhattan. His classes, lectures, and seminars on all phases of occultism were very well attended at the Warlock Shop/Magickal Childe Bookstore for over ten straight years. During that time, the Abbot witnessed the birth of the Wicca movement, the rebirth of the New York area OTO, and the formation of a band of artists, writers, musicians, and dancers—initiates all—called StarGroup One, which held monster promotional events featuring live bands and entertainment . . . all for the proselytization of Thelema and the gutsy anarcho-capitalism so well represented in the *Illuminatus!* trilogy by Wilson and Shea.

The Church never knew about the Abbot's extracurricular activities. It was essential that they didn't, for how else was he to get his hands on the *Necronomicon* and get it translated before the Feds found out about it? Not that the Abbot was a pious Christian. He believed that his ordination served him admirably in several ways: in the first place, it attuned him to the concept of the mystical transformation of matter and spirit that is the heart of the Divine Liturgy (what Catholics call the Mass). At the moment of transubstantiation, the Abbot felt at one with the ancient priests of Aton, the Gnostic adepts of Asia Minor, the Mithra cultists of the Middle East, and the worshippers of Attis in ancient Rome. In the second place, Ordination—and, with it, Apostolic Succession—transferred power of a Solar type. As priest-magicians of every age in the last two thousand years have always known, the combination of ordained priest and dedicated magician is a potent one. It is alchemy, sacrifice, resurrection, and transformation.

But the Abbot knew that the Old Age was over, knew it as surely as did the hordes of young revolutionaries who stormed through the City's streets in the late Sixties, protesting the War, the Government, Religion, and the inhumanity of Society. Surrounded by intrigue—both political and social—the Abbot came to realize that, as valid as the *rites* of transubstantiation might be, the ideology and moral precepts on which the Church was based were doomed: relics of an age long dead, confused and misinterpreted scriptures pasted together by self-serving bureaucrats hundreds of years after the death of Christ, all organized by a licentious leadership that laughed at the saints, that scoffed at revelation.

Magick was the only answer, the last key left in a Dead Age. Magick was the key, but where was the lock?

The Abbot, of course, was Simon. The lock, the *Necronomicon.*

The translation—carried out under secret and often exhausting circumstances—was completed on October 12, 1975: the 100th anniversary of the birth of Aleister Crowley, and dedicated to him. It would be over two years, however, before the *Necronomicon* was finally published.

In late 1976 it fell into the hands of one Lawrence K. Barnes, a gifted artist and scion of a family of printers and lithographers in New York. A serious devotee of the Cthulhu Mythos, L.K. Barnes had dreamed of the day when he would hold in his hands the fabled volume of occult lore known as the *Necronomicon.* Passing through Herman Slater's Magickal Childe Bookstore one day, he casually—jokingly—asked Herman if he had any copies of the *Necronomicon* laying around.

With a smile, Herman reached under the counter and pulled out the translated manuscript.

Herman Slater's influence on the translation and publication of the *Necronomicon* has often been ignored, and it should be set to rights here, once and for all. Simon had no idea what the *Necronomicon* was until Herman identified it for him. Simon had never read any of the Lovecraft opus prior to seeing the *Necronomicon*, and in fact disdained gothic horror and "occult" fiction altogether, much preferring the actual grimoires themselves to anything a pulp writer could dream up. Therefore, he had actually never heard of the *Necronomicon* until he laid his hands on it! As unbelievable as that sounds today, one must remember that in those days there were no other *Necronomicons* in print: the Giger book of artwork had not appeared, and L. Sprague de Camp's artistic concept had not been published, nor had the Colin Wilson/Robert Turner scholarly volume been printed. Certainly, Waite's books never mentioned it, nor did Francis Barrett's enormous volume, nor anything by Regardie or Crowley, nor the medieval grimoires themselves. The fact that the word *Necronomicon* appeared in the short stories by Lovecraft and in some comic books was completely unknown to Simon, who read neither. When he wasn't translating the All-Night Vigil Service of the Orthodox Church from Church Slavonic and Greek texts into English, he was making his own copies of *Le Dragon Rouge* and *La Poule Noire*.

It was Herman Slater who described the book's importance to Simon, Herman Slater who kept pressing Simon to finish the translation, and Herman Slater who kept his eye out for a likely publisher. Incredibly, several publishers turned the book down. (The book would later sell out three hardcover editions, and the paperback version—first published by Avon in 1980—has never been out of print a day.)

The *Necronomicon* was published, amid much fanfare, in December of 1977—the same year as the death of Elvis Presley, the rampage of the Son of Sam, the release of *Star Wars* and *Close Encounters of the Third Kind*, and the discovery of Chiron—in a deluxe, limited leatherbound edition of 666 copies, and a clothbound edition numbering 1333 copies. These were sold out within weeks, as many as two hundred of the leatherbound edition *months* before publication. A second edition was quickly printed, to be followed by a third edition.

* * *

In all that time, from 1977 to the present day, thousands of individuals have attempted to perform the rites described in the pages of that notorious, nondenominational grimoire. Some have been members of famous occult lodges and secret societies. Others have been individuals with years of private occult practice behind them. Still others have been people who had never before attempted an occult operation of any type.

And, increasingly, the *Necronomicon* has become associated with some of the sleaziest cultoids this side of a tabloid headline.

Fundamentalist Christian publications have damned the *Necronomicon* as some kind of Satanic bible, finding evidence of *Necro* activity in cult sites around the country. The *Necronomicon* has been linked to cult murders, kidnappings, insanity, and criminal activity of all kinds. Sigils from the *Necronomicon* have popped up in manuals for use by police officers investigating occult-related crimes. Preachers have denounced it from the pulpit; radio talk-show hosts have interviewed disturbed teens who said they had been turned into demonoids by the book. Heavy Metal rock bands sport *Necronomicon* symbols, and there was even a Japanese-made video game with *Necronomicon* sigils on it, called "The Wizard."

At an increasing rate, the *Necronomicon* has been linked in the media to LaVey's *Satanic Bible*, making the pair perform a kind of Devil's Duet through the pages of yellow journalism and supermarket tabloids.

Although Simon feels a great deal of respect for the basic thesis expressed in Mr. LaVey's books (and, anyway, they have the same publisher!), there is very little reason to associate the *Satanic Bible* with the *Necronomicon*. LaVey's system is one of breaking free of the guilt and repression that society uses to enslave its children. It is an ideological system: psychological shock treatment. Satan, in LaVey's eyes, represents freedom and individual Will. The *Necronomicon*, however, is less about ideology and more about a specific practice. Of course, there are demons aplenty in the *Necronomicon*, but no suggestion is made that one must worship them or even align oneself with them. Although being a Satanist, LaVey-style, does not disqualify one from performing *Necro* rituals, it is certainly not a requirement, either.

The rituals in the *Necronomicon* open a Gate. To use the *Necronomicon* is to take that first step from which there is no turning back: the step over the Threshold to the Other Side. Sometimes this step is accompanied by danger: danger to the psyche. To one who is *not* free—free from guilt and repression—the trip is all the more dangerous. Perhaps that is why so many readers have automatically associated the two books. While it is not necessary to worship Satan or to align oneself with the alleged Forces of Evil before attempting the *Necronomicon* workings, it *is* advisable to be a relatively secure human being, emotionally and psychologically if not spiritually. Unbalanced individuals toying with the rituals of the *Necronomicon* are, quite simply, doomed.

Groups of unbalanced individuals attempting to use the *Necronomicon* rituals as part of their group workings are on a suicide run. If a group has so little self-esteem that it must resort to slaughtering animals to build up their image, then they should definitely leave the *Necronomicon* out of it. For their sake, as well as for ours. The issue of blood in a *Necronomicon* ritual can have *extremely* unpleasant consequences.

And—although from my own extensive work in the occult community over the past thirty years, I do not credit the vast majority of reported "cult killings"—if *you* are one of those deluded, credulous types who believes that Pazuzu or Lucifer will love you more if you murder a human being in its honor: think again, fool. Simon is telling you now that a blood sacrifice of that type will cause Cthulhu to gleefully rip you apart—*from the inside*.

Therefore, I take this opportunity to personally disavow any misguided soul who creeps off into the woods to dismember a human in the name of the *Necronomicon*. You don't know what you are doing. The sacrifice will *not* bring you greater power. It *will* contribute to your own destruction. Should you persist in doing so, having now been warned by me, know that *thou art cursed from the very heart and soul of the Book.*

Blood sacrifice *should* be discussed, however, and readers are advised to pay special attention to this section as if their lives depended on it.

Blood sacrifice has been a part of religion for at least as long as recorded history. It is *still* part of many important religions, even today. It developed through an instinctive understanding that blood somehow carried the essence of life, of a human being, and that the spilling of blood meant that the ultimate commitment had been made, the ultimate sacrifice: for when blood was spilled, the creature died. Some cults believed that the soul of the slain creature carried the cult's message straight to the gods. Others believed that a bloody sacrifice was a sign that the cult was forever tied to its gods, bound in a bloody union. Still others believed that the spilling of blood called various supernatural beings to the area around the sacrifice, vampiric entities that could be trapped and made to serve the cult in any of a number of ways.

Now, with the prevalance of so many diseases that can be transmitted through the blood, we can appreciate blood as a potential killer, as well. What used to carry life, can also carry death. One would think that alone would be enough to stop cult killers from carrying out their deadly deeds: after all, one never knows if one's victim has any one of several diseases that can be transmitted from its spilled blood to another's open wound, perhaps. In the end, the only blood you *know* can't kill you is your own.

Yet, blood sacrifice can be achieved in other ways, as well. There is an entire sexual element to the sacrifice. We know that Crowley often referred to his own "sacrificing of children": a sly reference to the practice of masturbation. Indeed, the sexual rites are themselves blood rites, also. The bloody sacrifice of the Catholic Mass has its sexual analogue in the Gnostic Mass of Thelema, for instance.

While the author does not condemn the sacrifices that take place annually around the world among Muslims and Hindus, as well as among practitioners of Santeria, Macumba, Voudoun, and other Latin and African religions, he does insist that most individuals who are attempting occult work for the first time have absolutely no concept of how to properly perform such a sacrifice, nor do they fully understand the consequences. Therefore, the author strongly recommends an initiated course of study of several years' duration at the feet of an acknowledged master of one of the above-mentioned religions—someone who is experienced in both the theory and the practice of animal sacrifice—before ever attempting such a rite on one's own.

The Gates concept is central to the idea of the *Necronomicon* itself, not only in the book as edited by Simon, but also in the stories by H.P. Lovecraft that made the *Necronomicon* famous. In the stories, a "Gate" is opened to an extradimensional Place by someone using the *Necronomicon*, and Something Awful slithers through to be fought by the righteous—or just the lucky—who push it back through the Gate from whence it came. *All* occult ritual is actually concerned with piercing this division, this invisible barrier, that exists to separate reality and fantasy, consciousness and superconsciousness, actual and potential, Being and Nothingness, *tonal* and *nagual*. This division is called—in the jargon of the *Necronomicon*—the Gate. Actually, in ancient Sumeria (the setting for much of the *Necronomicon* ritual), it was also referred to as a Gate. Their word for Temple was BAR, and was depicted as a square building with doors on each of the four sides: the prototype of the mandala of the East. The Sumerian Temple was a place where human beings could reach through to a subtly perceived "Other Side" to contact the superhuman beings they knew to exist there; who had, according to the earliest written records of human civilization, given them that civilization and a promise to return.

The Elder Gods of Sumeria have kept their promise.

A word to Thelemites, Golden Dawn devotees, and others who are initiates of the Great Western Tradition in ceremonial magick:

Much has been written and discussed within occult groups and publications concerning the Enochian system of magick. Briefly, this refers to a complex series of magical squares, a magical alphabet and a complete magical language discovered by the Elizabethan magician John Dee and his mediumistic assistant, Edward Kelly. Discovered, that is, or invented, or revealed to them by angels. It was the Meric Causabon edition that was used by the Golden Dawn magician MacGregor Mathers in his famous synthesis of ceremonial magic, Elizabethan Enochiana, Qabala, and Egyptology.

Unfortunately, there were errors in the Causabon volume. These errors were compounded by a few misperceptions by Mathers and perpetuated by Israel Regardie. Although the resultant system, the Golden Dawn system of ceremonial magic, is unique, revolutionary, and beautiful, these flaws have become obstacles to many sincere persons who have tried to use the system for their own self-improvement and spiritual growth.

To remedy this situation, a number of groups decided to jettison the Golden Dawn system altogether and instead rely on the original Dee manuscripts—which are much more numerous and complex than even the obtuse Causabon book implies—thus making certain that they were working with the true system, free of the defects incorporated by Mathers and the other Golden Dawn initiates, including Aleister Crowley.

While the author applauds the efforts of those who pursue this method with all seriousness of intent, he cannot help but feel that they have thrown out the baby with the bathwater.

That Mathers' system is flawed is beyond doubt. As wonderful and startling as it was, it was still kludgy. Things were made to "fit" that simply don't, and I don't mean only the copying errors from the Dee material. This is no reflection on Mathers. Like that of any young computer scientist of the present age, his invention worked . . . to a point. The bugs had to be worked out, the whole system engineered better to perform as it should. The problem was, no one in the Golden Dawn possessed enough of an emotional distance from the material to be able to fix it. The Secret Chiefs concept robbed most of them of the necessary chutzpah to challenge what was perceived to be "received wisdom." In a sense, the reliance of Mathers and his colleagues on the fraud of the Secret Chiefs conspired, in the end, to hobble the system they had so painstakingly and cleverly devised.

Once the initiates had passed beyond a certain level in their theory and practice, the system seemed to short-circuit on them. As happens with so many occult lodges, the Golden Dawn fell apart into warring factions with charges and countercharges of incredible fantasy and excessively ornamental intrigues. Leadership, as always, was questioned, challenged, and fought over . . . but no one thought to seriously investigate the System itself, except from a purely ideological point of view. Waite broke off and became a Christian, of sorts. Crowley broke off and formed his own cult. Yeats retired to his poetry. Mathers, to madness and death. No one, in all that time, questioned the Tables of Correspondences and the concepts on which they were erected. What they questioned, and fought over, was politics. The problem did not lie in political matters, however. The System itself was flawed and, in the view of this author, dangerous especially to those most able to take advantage of it: that is, dangerous to those with the sensitivity and sincerity required to master the System and to become Magi. The author will not belabor the reader with his views on leadership and how occult lodges should be run, except to say that he believes there should be no "leadership" in a political sense at all. Rather, he believes in a revolving hierarchy of spiritual responsibilities, determined by calendar and not by vote or force or charisma. That said, let us investigate what all of this has to do with the *Necronomicon.*

That book refers to an occult calendar, which takes as its starting point the day the Great Bear hangs from its tail in the sky in the sign of Taurus, the Bull. That used to happen on midnight on April 30, the date we all know as Walpurgisnacht, the European equivalent of the American Halloween. The Gate was open on that day at that time, and on several other days computed from that day. In other words, the occult calendar in use by the workers of the *Necronomicon* system is based on an actual stellar event, and not on a paper calendar whose days and hours have come to have for us little meaning in this day and age: certainly, little relevance to the universe outside our communities. The Gate was not about Pagan or Christian or Jew or any other religious denomination. It was not about festivals to obscure deities or ritual reenactments of seed fertilization.

It was about an opening in the Universe.

A rent in the cosmic curtain.

A way Out for anyone on Earth, regardless of where they lived or what religion or race or language was theirs or their parents'. It exists in the firmament, like the Sun, Moon, and planets. Unlike the case with astrologers, however, there can be no debate over what zodiac to use (tropical or sidereal), or what house division to employ to segregate the planets and luminaries into their twelve stellar locations. Rather, this calendar is adjusted according to simple observation: when the Great Bear hangs from its tail in the sky. For those of us who live in cities or in areas cursed by smog and other forms of pollution that render the night sky invisible or opague, quick reference can be had to the astronomer's tables and to celestial mechanics. This we have done for you, in the Tables of the Bear that begin on page 229. These Tables provide nothing less than a true occult calendar, based on the actual motions of the stars relative to the motion of the earth, and to the earth around the Sun. For the Gate is opened at a moment in space, at a place in time. And at that time, and at no other.

Why that precise moment?

We don't know.

It may have something to do with the wobble of the earth on its axis, the flux of the geomagnetic field, the combined effects of solar and lunar variations. Magnetic storms. It could be a purely local effect; that is, peculiar to our solar system. Or it could be connected in some deeper way to the actions taking place on stars fifty light-years distant from earth.

It is a certainty, however, that it was chosen not at random but because of an observed event: that is, something had happened on that day or during that time, something so extraordinary that the position of the Great Bear was noted as the only salient characteristic in the environment. And when the same or similar extraordinary event took place the following year, or decade, or century—and on a completely different day—it was noted that, once again, it was the hour when the Great Bear had assumed the same celestial position.

Our quarter days are easily understood as the days of the equinoces and the solstices, with zodiacal attributions according to the "tropical" calendar, of course, and not the sidereal, which would be more reasonable. To some experts, the sidereal calendar is the only one that makes any degree of sense for an occultist to follow; everything else is nothing more than sentimentality masquerading as tradition. To others, the tropical makes sense if it is understood to be an arbitrary division of the ecliptic with the vernal equinox as its zero point; in this fashion, "Aries" is more of a philosophical concept, a metaphysical "place" that comprises the first thirty degrees from the equinox, than it is an actual place in the sky.

Although it is not the intention of the author to become

entangled in this controversy, suffice it to say that the *Necronomicon* insists on this difference: that the day of greatest importance in its calendar is *not* a traditional month and day, but the day and hour on which a specific astronomical event takes place. It requires the magician to be completely aware of the environment, completely in tune with the harmony of the spheres. It demands a greater degree of attention to the movement of *actual forces* in the created universe.

This is similar to yoga. In that discipline, one does not simply *act as if* one's posture is correct, one's breath rate is deep enough or measured enough, or one's kundalini is raised. One is either successful, or one is not. It is not subjective or "flexible," able to be modulated with mood or desire (or conferred by document, decree, or the laying on of hands). One has either raised the Serpent Power, or one hasn't. It doesn't matter what the calendar or anybody else says.

In ceremonial magick, much is made of planetary hours and planetary days. In the *Necronomicon* system, success is measured in other ways. One is not given the passwords to the various levels or Gates: one must discover them on one's own. And one must begin the operations at a specific time, and no other. No one can initiate someone else into the *Necronomicon* system: *it* initiates *you*. There is no room for demagoguery, fascist leadership, spiritual dictators. No one else has the power. Only you do. And the System. Beyond that, all is useless to you.

For this reason, the author believes that as a system of magic it is quite possibly the perfect tool.

PART ONE

The Mythos of the Gates

CHAPTER 1: Gate Symbolism

Readers already familiar with medieval ceremonial magick (and with the esoteric disciplines in general) will remember that there is a distinction made between the microcosm and the macrocosm. The microcosm is usually thought of as "our world," meaning our immediate frame of reference, our immediate perception of reality. The macrocosm is the world "out there": usually beyond our immediate understanding or perception. As magicians, we seek to enter the realm of the macrocosm *consciously*, in order to effect change on the microcosm. The macrocosm is the sphere where the planets move in endless rotations, presumably affecting our daily lives with their passage through the zodiac. The common person is at the mercy of the planetary tides; the astrologer attempts to time his or her actions in accordance with them. The magician attempts to neutralize the effects of certain tides and enhance the effects of others, thereby rewriting the natal and transit charts in accordance with will. The common person and the astrologer are passive observers of the macrocosm; the magician is an active participant in the machineries of joy. While such a role may be extremely attractive to those of us who must struggle by day to cope with the distant motions of the stars—"the slings and arrows of outrageous fortune"—there is a price to be paid before such a desire can pay off. There is a great deal that we do not know about the workings of the macrocosm:

much that takes place is invisible to ordinary eyes. Although the same rules govern both micro- and macrocosms, we know so little about either one that tampering with the machinery is dangerous—not only to ourselves but to those around us.

In India this danger is described as "practicing yoga without a guru": particularly kundalini and tantric yogas, which work on the autonomic nervous system. The autonomic nervous system (ANS) works independently of conscious thought. It regulates heartbeat, breath rate, peristaltic motion, and a host of other vital, daily functions of the body. To tamper with the workings of the ANS is obviously dangerous. We can't always *see* what we're dealing with. We can't always *predict* what will happen when we experiment.

The same is true in ceremonial magick. Whereas with yoga one enters the macrocosm through a variety of physical and mental techniques, in ceremonial magick we enter the macrocosm through ritual. The concepts are much the same; the outward form is all that really distinguishes yoga from magick. Indeed, magick has been called the "yoga of the West" by Francis King and other writers, and it certainly seems to appeal more to the Western mind-set. It is intellectual, and therefore Apollonian, but it employs various heightened states of conscious awareness—some close to ecstasy—and is therefore also Dionysian.

How to pass from one state to another? From conscious, everyday reality to the superconscious reality of magick? One needs a Gate, and the ability to pass through it.

The magick circle of the medieval European magicians was a place between the microcosm and the macrocosm. It was a stylized representation of a perfect World, complete with the sacred names, signs, and numbers that represented Perfect Unity—the goal of what C.G. Jung called the process of individuation. The magick circle is a mandala, similar to those of the East and probably adopted from the mandalas and yantras of India, and from the original glyph of a temple space, the BAR of ancient Sumeria.

Yet, inasmuch as the circle represents the individual magician's concept of what Perfect Unity (i.e., Godhead) is, the circle is his or her Gate into the macrocosm. It is not merely a symbol of unconscious cohesion, discovered in dreams or through depth analysis, as it is in Jungian psychology. Rather, it is a *dynamic* symbol, one that is consciously *used* to effect a result: the goal of individuation, but on a grander scale. As magicians, we understand what Jung refers to as individuation to be but a preliminary phase of the Great Work; the next phase is the identification of that individuated (microcosmic) Self with the Macrocosmic Self: what people in other times, other places, have called God.

Then why the danger? Why the many warnings, the guarded secrets, the organization of cults in which initiates are led through various tests in order to "prove their worthiness" before the secrets are revealed to them?

What's the big deal?

If we are speaking about a goal as seemingly tranquil as Unity with God—something that appears to be the goal of every priest, monk, nun, and ascetic we have ever met or heard about—what could be dangerous? Why not throw open the Gate to everyone, reveal the secret for all to see?

Because madness, delusion, and death await those foolish enough who rush in where angels fear to tread.

Remember what was said above about unguided ANS experiments. The danger lies not in the practices themselves, but in each individual person who approaches these practices unprepared. We do not know the extent to which our lives, our ancestors, our environment have damaged the delicate circuitry that makes up our individual organisms. We are not consciously aware of the inner workings of our glands, our nervous systems, or the gentle balance—or imbalance—that may exist between the play of our organs. We can stare forever at a map of the autonomic nervous system, for instance, but still not be able to know what is going on in our *own* bodies. We don't know what cells have been damaged, what nerves are impaired, the degree to which we can relinquish control of unconscious functions to the conscious mind. We can look upon our physical bodies as Gates to the other side . . . but we don't know if the hinges squeak until we try the door, and by then it might be too late.

In terms of our unconscious minds, we don't know what fears, complexes, fixations, even neuroses and psychoses, lurk there below our conscious understanding. Magick involves the whole person, conscious and unconscious, body and mind, every particle of our being. We would be hard put to catalogue every particle. There are "bugs" in our system: we need to be prepared to deal with them when they turn up. If we are not, the "bugs" become much more than tiny gremlins fouling up the works: they become demons, invisible Dunwich Horrors that can destroy us while the rest of the world stands by and watches, helplessly.

The preceding is a rather humanist approach to ceremonial magick. It is not the intention of the author to take away from those who firmly believe in the objective existence of demons, and of an actual Gate or Gates to other dimensions. The "humanist" approach is *also* superstition. It posits the existence of invisible mental constructs such as neuroses and psychoses, complexes, and something called the autonomic nervous system. These are all about as real to everyday humanity as demons or space creatures. They belong to the religion of the Twentieth Century. The venerable mythographer Mircea Eliade has written eloquently about the psychopathological states that would-be shamans enter; they only become shamans when—and if—they have cured themselves of their pathology. The fact that they have entered madness and survived as integrated persons indicates the extent of their shamanistic powers.[1] To the shaman, however, he or she is not having a "nervous breakdown." It is a possession by demons, a torture and dismemberment at the hands of supernatural forces. These are the images that contain the occult power and insight. They are the images of our dreams and nightmares.

Perhaps the symbols of psychology and medicine seem sensible and powerful to some; if so, then they are as valid as the medieval concepts of angels of light and darkness, fighting over the earth, using our souls as battlegrounds. It is all metaphor. What magick is about is practice: which action produces what effect. You can use your own ideological notation to describe these effects. For now, however, and in the following pages, the author will illustrate some of the notations used by other societies, other cultures than our own, to describe the same process: the Art of passing through the Gate. It is hoped that, in this way, every reader can find a suitable working metaphor to rely upon for the rest of his or her Initiatory Process.

The idea of a Gate is inseparable from the concept of the Underworld. To the Sumerians, there was no afterlife except in the Underworld. (That is, there was no idea of a Binary Underworld: Heaven and Hell, for instance.) Once dead—the most common method of entering the Underworld and, hence, of passing through its Gate—one spent eternity in a desolate landscape, drinking foul water. The only way to upgrade from Economy Class Underworld to Business Class was by dying a brave warrior. In that case, one spent eternity in the same desolate landscape, but at least one's water was slightly less foul.

One cannot help thinking that the priesthood of Sumeria had other ideas about the afterlife and a totally different concept of the Underworld. As this is central—not only to the *Necronomicon* but to *all* religions and cults—a little time should be spent in trying to figure out what the Underworld is, or is supposed to be, and why anyone would want to go there.

In counterpoint to the dreary descriptions of life after death for the average Sumerian, there is another mythos that is not verbally explicated but which is nonetheless obvious in the prayers, incantations, and religious iconography of that ancient culture. It includes the myth of the fish-being

Oannes as well as the ziggurats that are the hallmark of Sumerian civilization, as important to Sumer as the pyramids were to ancient Egypt . . . and possibly for reasons not altogether dissimilar.

The ziggurats were the main temple buildings of the seven principle cities of Ur. Each of these seven cities of Sumeria had its patron deity, and a ziggurat devoted to its worship. Worship, in this case, meant not only sacrifices and prayers. On certain days it also meant direct *communication.* The high priest of the local cult would ascend to a specially built chamber at the top of the ziggurat, alone, to enter into communication with the deity. There is no record to describe what actually went on in that chamber.

Yet, the insistence of so many Mesopotamian legends on the existence of a bizarre being called Oannes who led the Sumerians from a nomadic lifestyle to the sophisticated city-states they left behind leads us to believe that there was some time of extraordinary contact between the Sumerians and an alien Entity who jump-started their culture from wandering bedouin to a complex, urban civilization; from hunter-gatherers to farmers and merchants; from illiterate nomads to creators of poetry, myth, and ritual, to astronomers and priests, mathematicians and scribes. For the Sumerians, the Gate was opened from the *other side*, and when it did, it brought them only good fortune in the form of knowledge and enlightenment. For that reason, they erected huge temples from whose summits they would scan the heavens, searching for the gods who were promised to return, to lead them into the *next* stage of evolution.

For the Sumerians, this was a *collective* initiation they were seeking, the mass migration of their entire people into a future mode. After all, hadn't Oannes come—from the sky or from the sea, there is some confusion among scholars as to his actual point of origin, although everyone is agreed that he visited the Sumerians from somewhere under the waters—to bring enlightenment to the entire *race*?

However, for later civilizations this rush towards illumination took on a decidedly *individual* cast. Great numbers of people were no longer waiting for the arrival of an initiatory deity to perform the necessary magic: rather, individual persons of great courage and integrity went out to deliberately *search* for this missing god. Who knows what prompted this switch from a people united in its quest for illumination to the efforts of isolated individuals to find God? Was it the destruction of the Sumerian culture by the invading Assyrians that caused a general feeling of abandonment and hopelessness? One merely has to read over the common Sumerian lament—"Spirit of the Sky, Remember! Spirit of the Earth, Remember!"—to feel the poignancy of a race abandoned by its gods, a race that *knew* the gods existed, for it was a god itself that wrought the miracle of transformation of their own society into the *first* that knew writing, poetry, astronomy, mathematics, and priestcraft. The advance of the Assyrian armies and the sack of Babylon must have thrown the Sumerian intelligentsia into total, existential, despair.

Or perhaps the ruling classes saw in enlightenment and civilization a potent weapon with which to advance their own careers; something to be withheld from the general population and only doled out on occasions to those trusted enough to use the initiation in the service of the realm?

The author likes to think that some members of the Sumerian priesthood escaped the destruction of Ur, and fled to Egypt, perhaps, or to the Indus Valley, where they continued their search of the sky for the promised return of the gods. It would help to explain the curious shift in focus from a *group* waiting for God to the *individual hero's* Quest for god, a Quest that is composed in every mythos of the search for a Gate, its Opening, and the dangers inherent in the entire Process. It would help to explain the myth of Isis, searching throughout the Nile Valley for the scattered pieces of her husband and consort, Osiris. It would help explain the numerous Quests of Greek heroes such as Herakles and Jason. And it might even help to identify the source of the Star Gate beliefs of the ancient Chinese sages.

For the story of Oannes is an historical one, from the earliest recorded civilization in the world: the first society with a written language and a sense of its own ancestry. The cuneiform tablets of the Sumerians predate Chinese writing by over a thousand years, recording everything from astronomy to religion to medicine. We had a tendency in the last years of the Twentieth Century to look upon the myths of the ancient peoples as cartoons, fantasies used by the rulers of a society to entertain the populace, or to keep them subdued. The author submits that the myth of Oannes—so detailed and particular in its description of the entire event—is based on an actual communication between the people of Sumer and a being from another, alien culture. Whether that culture was extraterrestrial, as some have claimed, or purely earthbound, remains to be discovered. One thing is for certain, however: if the story of Oannes is true, and there is no "supernatural" explanation, then there existed *before* Sumeria another civilization—far more advanced—with writing, astronomy, medicine, agriculture, and mathematics. If so, what happened to *that* culture? Why is there no trace of it? And why did Oannes go to the trouble of educating the nomadic Sumerians? *Cui bono*?

It may be sobering to recall that, as this is being written, a scientist of the stature of Nobel Prize winner Dr. Francis Crick—who, with James D. Watson, discovered the double helix structure of the DNA molecule in 1953—still insists on the serious possibility of his theory of "directed panspermia": the idea that DNA or its primeval predecessor was "seeded" into our planet by aliens from another planet.[2] Taking that as a possibility, isn't it equally reasonable to assume that these same aliens would come back at a later date to check up on their "experiment"?

It is not the intention of the author to spin off into a Von Dännikenesque reverie about ancient spacemen. It is not necessary to believe in Oannes or in the belief systems of the Sumerians (or even in the wisdom of Dr. Crick!) in order to benefit greatly from the rituals of the *Necronomicon* (as we shall see). But it is tempting to believe that the mystical Gate we will open through ritual is somehow connected to a physical Gate between dimensions, a Gate that a mysterious half-man, half-fish creature named Oannes used to bring the light of science to our Sumerian ancestors some six thousand years ago, after waiting at least 100,000 years for Homo Sapiens to come up with it themselves. It is also tempting to believe that a Gate similar to that used by Oannes has been used by other people in other parts of the world since that time . . . and that the use of the Gate has not always been so one-sided.

Gates figure prominently in all religions. It implies, perhaps, a kind of resentment against a class-system that would erect buildings with Gates through which only the elite could pass. One imagines a nomad arriving from out of the desert and seeing before him Babylon in all its glory, only to find that he must pass through an opening in the city wall that is guarded by armed men. There are no gates in the desert. There is no such concept. Why, then, do humans build gates?

Gates imply access, viewed optimistically. On the other hand, they also imply *restricted* access. Secrecy. A Gatekeeper. A regulation determining when and if a Gate should be opened or closed. This creates—or represents—a kind of dichotomy in society between those outside the Gate and those inside the Gate. Two separate worlds—microcosm and macrocosm—but two worlds that communicate with each other: through the Gate.

Buildings with gates may also symbolize that most archetypal of gates: the womb. There exists a great deal of fanciful allegory concerning copulation, conception, pregnancy, and birth in the ancient cultures. The fact that blood issues forth from this Gate at periodic intervals was a source of great philosophical speculation among the adepts of the ancient cultures. Add to this the very periodicity of the menstrual cycle—so similar to the lunar cycle—and the fact that this monthly issue of blood ceases during pregnancy for another observable period (this time of nine months) and we have the basis for an entire mystery religion right there. A woman, a Gate, blood, a monthly cycle tied to the Moon, a nine month period during which the lunar cycle is somehow superceded, and the appearance of a new human being, which is also the announcement of the advent of another lunar cycle. Bury the Gate-born human in a hole in the ground when he or she has died, and you have yet another Gate. And you wonder about all the other humans who are also buried beneath the earth, and you imagine an entire necropolis of the dead, an Underworld.

This would have been fine, and would have gone a long way towards "explaining" the ancient mystery cults on a purely physical or biological level, were it not for the appearance of strange, otherworldly creatures such as Oannes who would descend from the sky or ascend from the sea, wearing strange clothing, making strange sounds, bringing enlightenment in the form of information about the stars . . . and then returning as they came, disappearing from view, forever.

Suddenly, the Underworld as merely a place where the dead go becomes a strained perception. There are two worlds, after all: a microcosm and a macrocosm. If there is a purely earthbound type of Underworld, then there is a macrocosmic—celestial—Underworld, as well. Oannes pointed to the stars. And returned to them. Maybe, in the course of conversations he had with the ancient Sumerian nomads, he used their own, limited, understanding of their world to explain deeper truths. He may have used elements of their "reality" as allegories to describe another reality.

And . . . he made the Sumerians look up.

Freud, towards the end of his life, became fascinated with what he perceived to be our common, instinctual urge *beyond* his famous pleasure principle, towards death itself. He noticed a tendency in some individuals to compulsively repeat unpleasant—i.e., un-*pleasurable*—experiences in an almost masochistic way. He pursued this thread of unfathomable behavior to its logical end, and came to the startling conclusion that a "death wish" exists in all humans to some extent, a yearning for an "inorganic state," as he expressed it. Although the evidence seemed to be there, the conclusion that humanity yearns towards death bothered Freud. It seemed to refute his basic premise that people acted according to the pleasure principle. The sexual instinct, he argued, was an urge towards life, not death. What, then, were these two warring impulses doing trapped in the same body?

Of course, many psychoanalysts had a hard time with the "death drive" concept, and preferred to think of it as some type of mystical theory of the Master's. They preferred a sex-linked theory of "aggression" as the bearer of all the hostile impulses in humans, and could not find a way to accommodate a death drive within the total structure of psychology.

After all, how could a human being harbor an unconscious desire for death—perceived by Freud and others as an "inorganic state"—unless the unconscious had some *prior experience* of this state? In other words, how can a basic component of the human psyche yearn for a state it has never experienced? The only alternative would be to admit that the psyche *has* prior experience of death: *personal* experience of *being dead*. This, science cannot admit. It implies a continuum of consciousness before birth and after death, and as science well knows, once the body dies, the individual dies and there is nothing remaining.

However, every religion, cult, sect—call them what you will—since the beginning of recorded history, has insisted on just such a continuum of consciousness. Of course, they may all be wrong. It may be nothing more than wishful thinking, a desire for immortality or a clinging to the concept of the beloved dead still existing somehow, happier in another form of existence.

Except: according to Freud, this "death drive" is a function of the super ego. Without going into a detailed discussion of Freudian psychology, let us simplify what is a fascinating theory by saying that the super ego acts as the transmitter of our higher culture and traditions. It is the entity that determines what we may—and may not—keep from our parental heritage: that is, what we *must* retain and what we *must not* retain from our earliest perceptions of our own parents. As such, the super ego is a function of the Oedipal complex (which is too familiar an idea to go into in detail at the moment).

What I wish to extract from this theory is a realization so stunning, so *enlightening*, that it is unfortunate it has not received greater attention among occultists, and that is: the unconscious operator that controls the transmission of human tradition from one generation to the next is the same operator that is responsible for the creation of humanity's higher *culture*, and is simultaneously the operator whose deepest, perhaps most visceral, impulse is the *death* drive.

If death were the end of existence, of consciousness, of personal identity, the foregoing would not make much sense at all. Yet, posit the existence of consciousness *after* death—a place to *go*—and the whole thing falls into place.

What is more: our entire proposal that there is a Gate between the microcosm and the macrocosm, a celestial Gate that confers immortality (a victory over "death"), becomes tenable and reasonable. We can interpret the Freudian theory to mean that our highest self—the self that produces music, art, literature, architecture, and, yes, religion—is also responsible for our unconscious drive towards death. If that is so, then the very idea of "death" must be reconsidered, and perhaps redefined.

Freud could not understand how the sex instinct (a life instinct) and a death instinct could cohabit the same psyche. They seem to be polar opposites. After all, one would think that sex/death is the perfect oxymoron. And, if death is the "inorganic" state (and Freud, as a scientist, knew well what he meant by "inorganic"), then it stands to reason that life must be "organic." Does consciousness transcend *both* of these states and, therefore, all of creation? Was Freud a pantheist in mufti?

Without going so far as to explain *why* these two instincts should coexist in the same psyche, consciousness engineers from China to India to the Middle East have explored ways in which the two instincts could become *united* in a single drive. They insist that the goal of each individual is not to remain at the mercy of these two, opposing drives (whose mutual antipathy, after all, may be the source of all aggression), but to unite them in a single impulse.

In a few short words, therefore, we have just described how the Gate is opened, and what it is.

The lay researcher only has to read through the accumulated writings of Mircea Eliade or Joseph Campbell to realize the extent to which spiritual enlightenment and initiation depends so strongly on experiencing death, even the terrible death of dismemberment and decapitation, by simulation. Initiates are led into dark caves, or tombs, or isolated buildings, and kept in total darkness for long periods of time during which they "die" or are "killed." During the time of "death" the initiates may undergo further experiences and teachings designed to enable them to cope with their new "reality." Then, the initiates are "reborn" and rejoin society.

The implied identification of the initiatory cave with both the tomb and the womb—with their associations of the death drive and the pleasure principle, respectively—is not accidental. And the identification of this experience of unity of the two drives with a celestial Pole and a celestial Gate is apparent in every mystery cult from the shamans of Siberia and Central Asia (who literally or symbolically climbed a tree or a pole to Heaven) to the adepts of ancient Egypt who oriented their temples and tombs with reference to, and gates pointing towards, the celestial Pole.

The mystery cults of ancient Greece were no exception, and their much discussed sexual rites and orgies expand our investigation into the realm of, not merely sexual symbolism, but even obscenity, or what some of us would call obscenity. From the myth of the sodomization of Horus by Set[3] to the Greek murals and vase paintings depicting every type of sexual union, to the practices of the Indian and Chinese sex magicians, it is not merely *sexuality* that is being incorporated into spiritual liberation, but the obscene in all its forms.

To the Greeks, the Goddess of Obscenity was also the Goddess of Death and of Witchcraft: Hecaté. The very word *Hecaté* means "bringer of light," equivalent to our Latin word *Lucifer*. Hecaté is often shown at a crossroads (a Gate being thus implied) accompanied by howling dogs and carrying a torch in her descent into the Underworld. Dogs, to the ancient Greeks, "suggested all sorts of obscenities."[4] Freudian analysts would understand this symbol quite well, for to them the Underworld is the Unconscious, and the Unconscious is the repository of all sorts of "shameful" or "indecent" material. Thus, it is not only the "dead" who are buried in the "Underworld," but also repressed sexual desires.

While these desires are often depicted as sodomy, bestiality, etc., the preeminent "indecent" desire is the wish for carnal love with one of one's parents. The Oedipal complex—or the "nuclear complex," as it was originally called—is the centerpiece of Freudian psychoanalysis. It's the ultimate taboo, and the source of much alienation and psychic discomfort. Yet, the initiate's return to a symbolic "womb" is nothing less than the enactment of a *fulfillment* of this deeply repressed—or surpressed—desire.

As Eliade writes:

> The same motif is documented among Western alchemists: the adept must return to his mother's breast, or even cohabit with her. According to Paracelsus, "he who would enter the Kingdom of God must first enter with his body into his mother and there die." Return to

the womb is sometimes presented as a form of incest with the mother.[5]

Although veiled, this impulse survives even today among the relatively modern (early Twentieth Century) rituals of Thelema, for example in the rite known as *Liber XXXVI, The Star Sapphire*. In this ritual, Latin incantations allude to a unity between Mother and Son and between Daughter and Father. In this same ritual, the magician is expected to make the signs of Set Triumphant and Baphomet, and there is the injunction that Set shall appear in the Circle.[6] The importance—the necessity—of Set, the Egyptian God of "evil," in this entire discussion will be made clear shortly.

What we should concentrate on now, however, is this twin aspect of mystical motherhood: at once giving birth and nurturing . . . and the tomb of every initiate. This implies a destructive womb and a constructive tomb.

Apart from the obvious sexual symbolism of a Gate as the Vagina, there is another level of Gate mythology that is related to this discussion. To psychologists and researchers in the humanities, this is known as the *Vagina Dentata,* or the Vagina with Teeth. This bizarre ikon of a devouring vagina with sharp incisors crops up in dreams, in art, and in some folk tales. According to Freudian analysts and others descended from that great Teacher, the Vagina Dentata is a symbol of castration anxiety, and they leave it at that. To an occultist, however, there is an even deeper side to this myth and one that is prevalent throughout the *Necronomicon* and other mystery/initiatory processes.

The idea that the Gate is *dangerous* and could destroy those who try to enter it is familiar to anyone who has seen the *Magic Flute* by Mozart or the many mandalas of India and Tibet that show fierce guardians at the gates. Even in Biblical mythology, there is an angel with a fiery sword at the Gate to Paradise.

There are—to social, civilized men—many vaginas free of teeth: those women not one's immediate relatives or otherwise ostracized by society. A dentate or otherwise dangerous vagina could only belong to a tabooed female, and the queen of tabooed females is the Mother, to be followed by the Sister and other women with whom sex would be considered incest or forbidden (such as very young, premenstrual, girls). One feels that the origin for these taboos may have been with Oannes, who, in giving the Sumerians civilization, may also have set down the sexual rules for that society as a necessary prerequisite for maintaining orderly lines of succession; in other words, the family unit. For a Biblical parallel, we need only refer to those extensive sexual laws given to Moses by his God, Jehovah.

Thus was the "common man" created. His guarantee of immortality was only through his offspring. As he died, he was reborn in his children (as, indeed, the DNA code is constantly reborn and is, thus, immortal). But there was a backlash to this imposition of culture and laws. The hero of this backlash was—in ancient Egypt—Set, the undying god.

Osiris, as darling of those who believed in life after death and the possibility of immortality in an afterworld, had died and was reborn. As archetype of the agricultural discoveries of the ancient peoples, he symbolized the grain that was cut down one season and planted in the ground the next (the Underworld), only to give birth to more grain the next season (Rebirth). Set, however, Set the Hunter, *never dies*. And that is immortality of a different stripe altogether.

Primitive beliefs concerning Gates and Doors—though often not identified precisely as such—can be located as distantly as the islands of the New Hebrides (now known as Vanuatu) in the megalithic societies of the island of Malekula, for instance. The Malekulans have been the subject of numerous books and essays by Dr. John Layard, Thomas Harrison, and A. Bernard Deacon among others. The geographical isolation of the Malekulans has permitted them to retain ancient rituals and belief systems relatively intact, and they have become the focus for much sociological research.

Without going into too much detail (interested readers can locate a copy of the Bollingen paperback *Spiritual Disciplines*, or *Stone Men of Malekula* by Dr. Layard), it will be enough to say that everything from human sacrifice to complicated labyrinth designs to stone dolmens can be found in aspects of this ancient cultural survival. What interests us at this time, however, is their belief concerning life after death and the Journey of the Dead.

Although the Malekulans believe in a beneficent deity who was the creator of humanity and the bestower of certain gifts—a being who, incidentally, arrived on the island one day by canoe and who lives "in the moon" and whose influence creates the spirits of all who are born—the deity they most concern themselves with, and fear, is the Devouring Ghost.

Depending on which part of Malekula one finds oneself, this Ghost is female, male, or of indeterminate gender. The Ghost will devour the soul of the departed unless it is propitiated with the sacrifice of tusked boars. When a person has died, their soul goes to a Cave where the Devouring Ghost waits. In other versions of the myth, the Ghost sits outside the entrance to the Cave and presents a puzzle to the soul in the form of a labyrinth that has been partially erased. The soul must finish the labyrinth before it can proceed past the Ghost and through the entrance to the Cave.

Here we have another version of the dangers inherent in passing through the Gate to immortality. If the soul cannot solve the labyrinth or, in other versions, offer the appropriate sacrifice, the soul is *devoured* by the Ghost. Immortality—in the form of life after death—is lost. The soul is consumed.

A symbol of this journey is the stone dolmen, which as Layard informs us, is the symbol at once of the Tomb, the Cave, and the Womb of a divine rebirth. The Cave is not the end of the journey. Rather, it is a passageway through which the soul must pass. A Gate, in other words.

In a part of the rite that is believed to be later than this first part concerning the Cave, the dead man then passes over or through a body of water to arrive at a volcano (in this case, a very specific volcano on the island of Ambrym). It seems as if the volcano has become the real destination of the soul, which, in some versions of the myth, dances on top of the crater all night until the morning star rises, at which time it falls down to await the night, when it can dance again.

An interesting aspect of this entire belief system is the idea of the *ta-mat* (a word that sounds suggestively like the Sumerian word for the primordial, underworld serpent, Tiamat). Although it means "dead man" in a literal translation, it is also a concept similar to another Melanesian idea: *mana*. To the Malekulans, the *ta-mat* is the accumulated psychic energy or life-force of an individual, which has been increased during life in a process called *ra-mat*, which means "to make a soul." According to Malekulan belief, an individual must make a soul while still alive or else there will be no part of that individual to survive death. Once an individual has reached that stage in life where—after a record of numerous sacrifices performed and other sacred deeds—it is believed he has accumulated a sufficient store of *ta-mat*, he may perform part of his own mortuary ritual in advance, signifying that he is already, symbolically, dead.

This is a sophisticated theme that we will come to again and again during the course of this study, particularly as it is reminiscent of certain Egyptian ideas concerning the *ka*. Suffice it to say for now that Biblical scholars of the stature of S. H. Hooke (mentioned in the original *Necronomicon*) were so startled by the similarity of the Malekulan rituals and beliefs to those of ancient Egypt and Mesopotamia that it has become the basis for several learned papers. The possibility that the rites of Sumeria and Babylon made their way to a group of islands off the coast of Australia and New Caledonia—by way of South India—is taken seriously today by modern scholars. The discoveries of Dr. Barry Fell of Harvard and other epigraphers have gone a long way to suggest that many ancient peoples were sufficiently experienced in sea travel to have accomplished such amazing feats of exploration centuries—if not millennia—before Columbus's conquest of the Atlantic. The parallels between the Malekulan myth of the beneficent "white-skinned" god who created humanity, who lives "in the moon," and who brought the islanders their first animals and foodstuffs (and, presumably, their religion and culture), to the Oannes myth of Sumer, suggests further fields of study. The idea that the dead proceed on a journey through a Cave, past a Ghost, and thence to a volcano, seems too similar to many Sumerian ideas to be totally accidental: is the ziggurat, the stepped pyramid of ancient Sumer with the sacrifical fire burning at the top, merely a symbol of the volcano? Or was the volcano a convenient metaphor for the pyramids of Ur, which led the people into communion with their gods?

Although the exploration of similarities between this South Pacific culture and the lost civilization of Sumeria is tempting, there is simply insufficient data on which to decide whether there was any contact between these two groups. However, the concept that holds the greatest interest for us now is whether Sumerian—or, possibly, Babylonian or Assyrian—missionaries or tradesmen made it as far as the South Pacific in ancient times. The theory has it that the similarities in ritual between the cults of South India to those of Mesopotamia (and to Malekula) are so great as to force us to consider whether such a migration of people and/or ideas did take place. If so, it would certainly give rise to speculation that the mysteriously appearing and disappearing Sumerians were the ones to "seed" the basic myths and rituals of humanity around the globe. And this, after their "jump-start" at the hands of the alien, Oannes.

Virgil recounts—in the Sixth Book of the *Aeneid*—how Aeneas descends into the Underworld through a similar Gate, this time guarded by a prophetess, the Sybil, who demands prayers and the offering of a golden bough (which is later referred to as a "fatal rod": *fatalis virgae*). The Malekulan deceased is buried with a rod cut to his exact height, which he takes with him on his Journey to the Underworld, and which he uses to part the waters of death in much the same fashion as Moses parted the Red Sea. In fact, quite a bit is made of this frightening entrance to the Other World in the *Aeneid*, complete with references to the Minotaur and other labyrinth motifs. I am afraid we must assume that the Malekulans were not conversant with the Latin of Virgil's *Aeneid* or the translations published since then; therefore, the similarity of these two myths must be based on a third, more ancient, reference source . . . or on a common, perhaps psychological, denominator in all races.

The problem with the psychological explanation—that humans share a common Ur-myth based on their simply being members of the same species—is that scientifically minded individuals have a hard time accepting the "ancient racial memory" or "collective unconscious" theories. Some believe it is actually *more* scientific to believe in a common, historical root for these various myths. That this approach may lead the "scientifically minded" back along a track that involves Sumeria, Oannes, and the possibility of alien intervention in earthly affairs seems a distant enough possibility not to bother the mainstream historian or scientist, "directed panspermia" and all its implications notwithstanding.

For our purpose, either point of view will do. Whether the origin of these myths lies in a collective unconscious or an ancient racial memory or the actual migration of Sumerian priest-magi to all parts of the globe is a matter for speculation. What *does* concern us is whether—either in the psyche or in the stars—there is a Gate and where it leads and how to enter it with our eyes open. And the only way to answer all of these questions is to examine the myths and rites of the peoples who professed to know all about it.

Frazer, in his classic work *The Golden Bough*, referred to many superstitions regarding locks and doors. These beliefs concern both birth and death, and involve locking and unlocking doors, gates, drawers, knots, and anything else that could be tied and untied.

The beliefs of people as remote from each other as the Germans of Transylvania, the Indians of Bombay, the Hos of West Africa, and the villagers of Scotland are alike in that it is believed that—in times of childbirth, particularly in a difficult birth—all doors, windows, and even drawers should be unlocked and left to swing open to permit the safe birth of the child. That is, nothing should be bound or tied or locked, so the spirit or soul of the child can freely enter and the passage of life into the world ensured.

This is echoed in the beliefs concerning death, which are the same: when someone is lingering on the brink of death for a long time and in terrible pain, doors and windows are unlocked and knots untied in a gesture of sympathetic magic so the spirit can be free to leave the body quickly.

These beliefs even extend to personal posture. In cases of pregnancy, the husband of the expectant mother is not permitted—in some societies—to cross his legs or to fold his hands. Conversely, a popular method of cursing someone in many cultures all over the world is known as "ligature." This involves tying a knot in something, a human hair for example, that represents the person to whom evil is being wished. This act of sympathetic magic is designed to bind the other person's vital energy—their *ra-mats*, *mana*, *prana*, etc.—causing them to sicken and die (at worst) or to be simply unable to harm others (at best).

This seeming digression is intended to demonstrate the universal belief in the action of vital energy (*mana*, *ta-mat, ch'i*, *jing*, *vril*, *soul*, the ancient Egyptian *hike* or *ka* or by whatever name it is known), its extracorporeality, and its relationship to the Gate and the Underworld. Many more examples could be given from ancient mythic sources, but we will satisfy ourselves for now with these. In the course of the following pages, other examples will appear and soon the premise of this book should become quite clear.

In the interests of saving space and of moving the story

along quickly, passing references will be made to many diverse cultures and religiomagical practices in the following pages. Rather than burden the text—and the reader—with many cumbersome footnotes and intratextual citations, I propose that the reader consult the bibliography at the end of this book, which is arranged by subject and contains suggestions as to specific fields of study. All research that has been conducted concerning the core theory of this book is based on primary sources, in so far as is possible, and on some very reliable secondary sources as well. *None* is based on automatic writing, channeling, astral travel, visions, etc. One of the main criticisms of Madame Blavatsky's otherwise excellent and compendious research was that she had a tendency to fill in the gaps in her theories with references to books or sources that were, shall we say, etheric. This has not been the case with *Gates of the Necronomicon.*

The reader may end up with more questions than answers at the end of this book. He or she may desire to check on my references, motivated by disbelief in the conclusions that are drawn, by horror in the direction to which they point. I urge the reader to do so. There is nothing more important to one's life than the complex question of one's own death and immortality.

CHAPTER 2: The Celestial Gate in Ancient Civilizations

The place is Egypt. The site of the soon-to-be-built Temple of Denderah. The time, a thousand years before the birth of Christ. The Pharaoh is about to build a new temple to the Gods, by establishing its four corners in a ritual known as "the Stretching of the Cord," peculiar to the Goddess Seshat, consort of Thoth.

The Pharaoh speaks:

> My eye is fixed on the Bull's Thigh Constellation. I count off time, scrutinize the clock, and establish the corners of Thy temple.[1]

The same country, some years later. The Pharaoh has died. As part of Egypt's famous funerary rituals, the priest touches the mouth of the Pharaoh with an instrument identical in shape to the constellation that was used to orient the temple. This ceremony is called "the Opening of the Mouth." By so doing, the priest restores life to the *ka*—or spiritual double—of the deceased.

The instrument—generally described as an adze or as a pair of adzes—is called by the same name as the constellation (actually, an asterism by modern standards) that was so important to the orientation of the temple: the Bull's Thigh.

A more common—and revealing—name for this same constellation is the "Thigh of Set."

Seshat, Goddess of Writing and Record-Keeping, Lady of the Builder's Measure, Founder of Architecture, Goddess of Construction, and Lady of the House of Books,[2] was also the Foundress of Temples, who was occupied with "helping the king to determine the axis of a new sanctuary by the aid of the stars, and marking out the four corners of the edifice with stakes."[3] She is depicted oddly. Arrayed in her leopard skin, she is crowned with a straight rod balanced upright on her head, which ends in a seven-pointed star at the top. This star is then surmounted by an inverted crescent, or pair of horns.

By our rather shallow contemporary standards, this specific Goddess is not particularly sexy, regardless of the leopard skin. She is, after all, a glorified librarian and part-time surveyor. One imagines her wearing thick glasses and carrying a stenographer's pad. She has not attracted much attention in the literatures of either Egyptology or occultism. At best, she's "Mrs. Thoth."

This general understatement of her importance has been successful in disguising one of the most important deities in the ancient Egyptian pantheon; very probably, Seshat holds the key to the inner Egyptian mysteries: the unwritten traditions that were behind the erection and orientation of the pyramids and the entire, obsessive nature of Egypt's Cult of Death. She is also the ancient forerunner of an entire tradition of Masonic and Western occult ritual and legend. Of Seshat we would say today: she forgot more than Solomon ever knew.

Her crown tells some of the story. The iron adze—the Thigh of Set—tells the rest. But to fully appreciate this story and all of its implications, let us move to pre-Columbian Mexico. The constellation known to the ancient Egyptians as the "Thigh of Set" was known to the Aztecs as *Tezcatlipoca*, a God who limped due to a fight with a sea monster in which he lost a foot, which was replaced by an obsidian mirror, hence the meaning of his name: "God of the

Smoking Mirror." Tezcatlipoca was an evil god, of whom it has been written:

> Mankind was simply forced to coexist with this awesome being. It is unusual to find a nation devoted to the service of a demiurge whom we, in a European tradition, would regard as evil in his innermost nature. The only possible parallel is among some of the early dynastic Egyptian kings, who worshipped Set, the spirit of the desert and its terrors.[4]

Indeed. But let us not stop here. Rather, let us turn now to ancient China. The Han Dynasty, perhaps. About 200 B.C. There, this same constellation is inscribed on the famous diviners' plates of that period: plates that also, at times, had small, magnetic, adze-shaped devices placed in the center. The Chinese called this same constellation by a wide variety of names: Winding Constellation, Northern Ladle, Northern Bushel, Emperor's Chariot, and, most important to us, the Celestial Gate.

It has been called "the pivot of Taoist astral mythology"[5] and "at the same time a place of origin and of return."[6]

A place of *origin*. And of *return*.

It is worshipped—invoked—by an awkward, labyrinthine dance called the "Pace of Yü."

It is nothing less than the familiar asterism of modern times: the Big Dipper.

What we in America call the Big Dipper—and in England, the Plough—is really part of a greater constellation called Ursa Major, or the Great Bear. This is the constellation referred to in the *Necronomicon* when the advice is given that the Gate can be opened when "the Great Bear hangs from its tail in the sky." Its reputation is homely. It is the easiest group of stars to locate in the northern latitudes, and every schoolchild can find it. It is so familiar, so ordinary, that we tend to forget its tremendous importance to early navigators, astronomers, and high priests.

For us in the northern latitudes, the Dipper is circumpolar. That is, it circles the North Star and is never below the horizon. The ancients were able to tell time from the position of the Dipper, as it appears to rotate once around the North Star every twenty-four hours, give or take a few minutes.

Those "few minutes"—actually a little less than four minutes per solar day—can add up from day to day, so that at midnight in June the Dipper will point in a completely opposite direction from midnight in December. No matter. If you knew what month it was, you could tell the time by finding the Dipper at night.

Another handy feature of the Dipper is the fact that two of its stars can be counted on to point straight at the North Star, Polaris, thus giving mariners a sense of direction in the midst of the featureless sea.

But, to the ancient people, the Dipper was much more than a kind of macrocosmic clock and pointer. It had something to do with Death, with Enlightenment, with Astral Travel, and with the worship of their Gods. It was a Gate, the passageway to heaven:

> . . . the Dipper represents the gate of access, the passageway. . . . certain meditation methods involving the Dipper are called the Veneration of the Seven Stars Which Allow Passage. The seven passages refer to the pace of the Dipper, the seven stars are at the threshold of the Gates of Heaven. The last star of the Dipper is also known as the Celestial Gate, and sometimes the constellation as a whole is given that name. The lords of the Dipper give the practitioner the "Talisman that opens the gate."[7]

This "talisman," of course, is well known to readers of the *Necronomicon,* for it was just such a talisman that enabled the Mad Arab to open his first Gate.

This is a lost knowledge, a forgotten tradition, whose residue has nonetheless survived in fragments of ritual and in cryptic allusions to "astral bodies" and "seven planes," "seven chakras," etc. The Golden Dawn's "Lesser Banishing Ritual of the Pentagram" is a manifestation of this tradition, as is the Thelemic "Star Ruby" ritual. As is the "Middle Pillar Ritual" of the Golden Dawn. Indeed, the casting of the Circle in Wiccan ceremonies as well as the elaborate conjuring mandalas of the medieval sorcerers of Europe and the mesmerizing tangkas of Tibet, Nepal, and Gansu Province in China are all survivals of this primeval science. It's the missing link in hundreds of cults, sects, and religions the world over. It makes sense of many of the rituals and beliefs of groups as diverse as the Daoists of China; the Aztecs of Mexico; the Witches of Europe; the Ceremonial Magicians of Europe, Africa, and America; the Golden Dawn; and Crowley's A..A..; it's both their common denominator and their greatest secret.

And, for us, it begins with the *Necronomicon*.

At first glance the initiatory framework of the *Necronomicon* seems to fit the mold of many mystery cults—seven gates, corresponding to the seven "planets" of the ancients: Moon, Mercury, Sun, Venus, Mars, Jupiter, and Saturn. These are referred to as the *Zonei*: a Greek term meaning the "Zoned Ones," that is, those traveling in their own particular zones or orbits in the sky. Those that are "unzoned" are the fixed stars, since they seem to be stationary in the sky relative to the motion of the planets, the Sun, and the Moon. Generally, the fixed star constellations that preoccupy most astrologers, occultists, etc., are those that make up the Zodiac. The Zodiac, as we all know, is a belt of twelve constellations against which the planets and the luminaries appear to move.

However, there are many visible stars in the night sky that are not part of this zodiacal belt at all. Orion is one of these, as is Cassiopeia, Canis Major, and Canis Minor, to name but a very few. Perhaps the single most notable constellation in the northern latitudes, however, is the one we all know as the Great Bear, Ursa Major, and its component asterism, the Big Dipper.

Since the planets never "enter" these constellations, they are generally unimportant to astrologers. They just "sit there," revolving endlessly around the celestial pole. The constellations that never set below the horizon are called "circumpolar," because they make a grand circle around the pole. In our era, the celestial north pole is a spot in space very close to the star called Polaris, or the North Star. But in previous eras the "pole star" was a different celestial body. In 3000 B.C. it was the star Thuban in the constellation known as *Draco,* or the Dragon.

The *Necronomicon* makes mention of this, on page 208 in the second part of the Mad Arab's testimony:

> And there shall forever be War between us and the Race of Draconis, for the Race of Draconis was ever powerful in ancient times, when the first temples were built in MAGAN, and they drew down much strength from the stars . . .[8]

And again on page 210:

> They worship when that Star is highest in the heavens, and is of the Sphere of the IGIGI, as are the Stars of the Dog and the Goat . . .[9]

And in other places in the book. This would seem to be a reference to those ancient times—five millennia ago—when Draconis was the most important constellation and seemed to pivot around the heavens on one of its own stars, Thuban (Alpha Draconis). The gradual change from Draconis to Dipper seems to signify—to the Mad Arab—a change from a world run by a bloodthirsty cult of Serpent worshippers to one run by a more orderly and benign civilization. Indeed, to the ancient Egyptians the Serpent symbolized the "Most Ancient One," the First God, who dwells beneath the waters of the earth and waits for the time when it will rise up again and destroy the world as we know it. This act of destruction will unite "the waters below with the waters above"[10] in a state equivalent to that primordial epoch before the creation of humanity and its subsequent trashing of the cosmos. This entire idea—of a sea monster dwelling beneath the seas, dead but dreaming, only waiting for a chance to rise up and destroy humanity—is the archetype of Lovecraft's Cthulhu Mythos, virtually in every significant detail, even to the concept of the Ancient Ones. The Egyptians of old would have been at home with Lovecraft's stories, as would the priests of the vanished race of Sumeria.

But while it is tempting to assign political or theological content to the story of the war between the Cult of Draconis and the relatively present-day Cults of the Bear, it is wise to remind ourselves that this shift from Draco to Polaris is very much a function of such things as gravity and the combined gravitational pulls of the Sun and the Moon, which cause the earth to wobble on its axis in that eternal dance known to us as the "precession of the equinoxes." For the shift from Draco to Polaris is but a circumpolar manifestation of the zodiacal shift from Aries to Pisces to Aquarius as the equinoctial point of the New Year.

While every student of astrology knows that when the Sun enters the first degree of Aries the zodiacal year begins (i.e., at the Vernal Equinox), what is not so well understood is that the Sun has not actually "entered" the first degree of Aries on March 21 for thousands of years. For reasons it would be too technical to go into here, the wobble of the earth on its axis means that the celestial poles shift around a giant circle in the sky approximately once every 26,000 years. In India, as in other places where Sidereal Astrology is practiced, care is taken to ensure that the astrological charts are drawn for the "actual" positions of the Sun, Moon, and planets in the heavens as opposed to the traditional positions used by Western astrologers in what is referred to as the Tropical Zodiac. That means, for instance, that a person born on March 21 in 1950 was not born in Aries at all, but in early Pisces. While this may seem to throw current Western astrology into chaos, it is actually a way to remain conscious of the *actual state* of the heavens and the earth's angle and relationship to the heavens at any given time. It reminds us that the heavens are constantly in motion, and that nothing sits still for long. Not even the Zodiac.

What this means for an occultist or magician is crucial to an understanding, and more effective use, of the forces that surround the magick circle. It permits the magician to approach the Celestial Gate directly. It aligns the magician with the actual forces at work in the cosmos, rather than with a simulacrum of those forces. In short, it makes the Work easier, and more potent.

Everyone is familiar with the expression "Age of Aquarius": a new age in history that will bring about a general enlightenment of humanity, an age of spiritual liberation. Not everyone is aware of just where the phrase "Age of Aquarius" comes from. Due to this phenomenon of the precession of the equinoxes, the Vernal Equinox of March 21 will take place—not in Aries or in Pisces—but in Aquarius. The date when this will occur is the subject of some debate, but most astrologers agree that it will happen early in the third millennium: that is, shortly after the year 2000 A.D. That means, of course, that as this book is being written we are in the last days of the Piscean Age, an age that was ushered in by Christianity and the military, economic, and religious dominance of Western, European nations over the lands of Asia, Africa, and native America that has been Christianity's legacy to the world.

To other occultists, the Age of Aquarius—although it is a phenomenon of astronomy and astrology—corresponds to another New Age, the Age (or Aeon) of Horus. Traditionally, this age began in April 1904 when Aleister Crowley received a series of communications over three days in Cairo from an incorporeal ("inorganic"?) entity called Aiwaz. The message from Aiwaz concerned the end of the old spirituality represented by the Buddha, Christ, Moses, and Muhammed, and the beginning of a new age symbolized by the ancient Egyptian deity, Horus. Some believe that this age will trigger a Third World—Southern, Eastern—dominance over the Northern, Western, nations of Europe and North America. This will become important to us later on, and is mentioned here only to prepare the reader for some of the information that will follow. The question that will concern us is whether a Gate was opened those three days in Cairo in 1904. If so, what type of Gate was it? And what does it mean for the magician of the *Necronomicon*?

Before we answer these questions, let us investigate the Celestial Gate itself: the Ursa Major constellation as it was revered among the ancient peoples of Africa, Asia, and South America. Then we will discover why the Mad Arab feared certain days when this constellation hung "from its tail in the sky." We will identify those days, and find out what are "the days computed from that day."

The Seven Rishis

The Dipper—or Celestial Gate—was uniformly regarded by the peoples of antiquity as the domain of beings who were *already* ancient, long before the founding of Pharaonic Egypt or Vedic India or Daoist China. In India the seven stars of the Dipper were considered the astral bodies of the "seven Rishis" or seers, the progenitors of humanity who descended from the mind of Brahma. This makes them incredibly ancient, particularly as the great sages of India compute time in such enormously long cycles.

> The Seers are mysterious beings connected with the origin of man and the origin of knowledge. Although represented as human sages, they are considered eternal powers who appear every time a new revelation is

> needed. . . . The seven main seers dwell in the sky as the seven stars of the Great Bear. They are connected with the divinities of the elements.[11]

These Seven Seers are cosmic principles. Each is named and given the appropriate wife or wives in the great cosmogonic scheme of things. They are as follows:

The Seer Light (*Marici*) married to Fitness (*Sambhuti*). Light is the Father of Vision, known as *Kasyapa*. This latter deity will become interesting a bit later.

The Seer Devourer (*Atri*) is wed to Without Spite (*Anasuya*). The Devourer represents the power of Detachment. He is also a law-giver, and is said to have given birth to the Moon from his eyes.

The Seer Fiery (*Angiras*) is wed to Modesty (*Lajja*). This Seer symbolizes the power of Illumination, and it teaches transcendent knowledge. Interestingly, he has four daughters by another wife (*Sraddha*, or Devotion) who represent four aspects of the lunar cycle: the first day of the New Moon, the New Moon day itself, the last day before the Full Moon, and the Full Moon day itself. *Angiras* is sometimes identified with Jupiter, and with all luminous objects. (This mysterious identification of Dipper stars with the planets is a theme that crops up everywhere. Also worthy of note is the insistence in the *Necronomicon* on the *thirteenth* day of the Moon as the day to begin the ceremony of opening the Gates, the ritual of the Walking. The thirteenth day can be considered the day *before* the Full Moon, and therefore corresponds to one of the four daughters of Angiras, *Anumati*, and the day when "the gods and Ancestors receive oblations with favor.")[12]

The Seer "Bridger of Space" (*Pulaha*) is wed to Forgiveness (*Ksama*). Pulaha is one of the Seers of the Antigods (*asuras*). These beings will be discussed below.

The Seer Inspiration (*Kratu*) is wed to Humility (*Sannati*). This Seer represents the power of Intelligence. Intriguingly, Mircea Eliade identifies Kratu with the heightened state of ecstatic fury typical of shamanistic initiations and cults, such as those of the *berserker*. The very term "berserker" means, literally, "*bear* shirt." The warriors who went about crazed on their murderous missions wore shirts or tunics of bear hide to denote their membership in the cult. This may also have had something to do with the werewolf legends that are common in many parts of the world. Also, might there be an etymological link between the Sanskrit *kratu* and the Greek word *krato,* meaning "power" or "strength"? *Kratos* is a word we find used in the classics to mean "a lust for power," a connotation quite similar to that divine madness implied by the Sanskrit *kratu*.

The Seer "Smooth Hair" (*Pulastya*) is wed to Love (*Priti)* and is a non-Aryan being. Remember that these Seven Rishis are the progenitors of the entire human race. Pulastya is the other Being who, with *Pulaha*, are the Seers of the Antigods, the *asuras*, of which more below. The Lord of the Antigods—*Kubera*—is the son of Pulastya, and is also the regent of the North cardinal direction.

The Seer "Owner of Wealth" (*Vasistha*) is rather cynically married to Faithfulness (*Arundhati*). His power is that of the Spheres of Existence, referred to in Sanskrit as *Vasus*. As wealth is necessary for the performance of the ritual sacrifices, Vasistha is also intimately connected to the concept of sacrifice: the offering up of existence or tokens of existence, symbols of the ego.

These, then, are the Seven Rishis and the Seven Stars of the Big Dipper or Celestial Gate, according to Hindu tradition.

"The stability of the world results from the rituals performed thrice daily, at dawn, midday, and sunset, by the seven seers . . ."[13]

Oddly enough, midnight is not mentioned, a point we shall return to later.

For now, let's look at some of the more intriguing aspects

of the Seven Rishis, and start with *Kasyapa* the Lord of Vision, son of the Rishi *Marici* or Light.

Kasyapa was the father of a bizarre Serpent God, *Sesha*, whose name means "Remainder." This Serpent—like so many other serpent gods around the world—dwells beneath the earth and is regarded as a primordial being, an entity that preexisted humanity and was present at the creation. Sesha is the Remainder of previous universes, which are destroyed "when creation recoils upon itself."[14] This god is disconcertingly similar in content to the ancient Egyptian Serpent God who dwells beneath the earth and is called *Wer*, "the Most Ancient One," and who comes from a primordial time before the creation of humanity and waits for the day when creation will be destroyed and the Serpent rules once again.[15] It is also eerily reminiscent of the Qabalistic tradition that the *qlippoth*, the "shells," are the demonic remainders of previous attempts at creation, the shards of the Sphere of Mars, which was not strong enough to contain the Divine Light and therefore shattered into myriad fragments.

Of course, the very name *Sesha* called to mind the Egyptian Goddess, *Seshat*, but I could find nothing particular to link the two *directly*. Indirectly, however, there were some circumstantial similarities that may or may not reveal an underlying identity.

For instance, Sesha is a serpent of a thousand heads, but is shown in carvings and ikons as having only seven heads. Is the straight rod surmounted by a seven-rayed star that is the crown of Seshat a stylized representation of a Serpent with Seven Heads, or Sesha? Generally, among the ancient Egyptians, a serpent is a serpent; there is no stylization taking place. But what if the concept were a foreign one, something borrowed in symbolic form from another place?

Another connection is the idea of great age and antiquity. Seshat, as the Goddess most concerned with orienting temples in alignment with the Celestial Gate—the area of the heavens where, according to the ancient Indians and, as we shall see, the ancient Chinese as well, humanity originated (remember Dr. Crick's "directed panspermia" theory)—she is concerned with primordial origins. Further, the Ancient Serpent in Egyptian mythology is also Apophis, the great enemy of Osiris, sometimes identified with Set; and the seven-starred Dipper constellation is sacred to Set.

So, Seshat is leading us to forge a link between Sesha of the Hindus and the Ancient Serpent of the Egyptians. Sesha, as mentioned, was present at the Creation. The gods of India wrapped Sesha around the mystical center of the universe—the provocatively named Mount Sumeru—and used him thereby to churn the celestial oceans. Mount Sumeru is regarded by many writers as the *axis mundi*, the axis of the world, which ends in the north and south poles. The Ancient Serpent, Wer, of the Egyptians is also called *Mehen* or the Encircler.

One of Sesha's incarnations is as *Bala-Rama*, a strong and rather fierce god, addicted to wine and a brother to Krishna. This incarnation is depicted as dressed in blue with a white necklace, and carrying a *plough*. My humble proposal is that a god dressed in blue—the color of the sky—and wearing a white necklace (the circumpolar stars?) and carrying a plough (symbol of the Dipper in some cultures and of the Pentagram in others), is an ancient representation of the Dipper. Whether this is true or not is open to verification by researchers more able in Asian languages and mythology than myself. What is pertinent to this investigation, however, is the *wildness* of Bala-Rama, his eagerness to fight (even with his brother, Krishna), and his addiction to alcohol. These are all attributes that can be assigned, with ease, to the Egyptian god Set.

One of the other Rishis, Pulastya, was the father of Kubera, Lord of the Antigods or Asuras. Kubera is a god of treasure buried underground, the precious stones and metals hidden in the earth. He is also lord of the *yakshas*, the "mysterious ones" and of the *guhyakas*, the "secret ones." Sometimes these two groups of demonic beings are considered one and the same. They are fellow travelers of the *rak-*

shasas, who are "night wanderers," and of the serpent demons, the famous *nagas*. In this way, Kubera is the King of Demons, what some authors might consider "earth elementals," particularly as their sole responsibility is the protection of the buried treasures of the earth.

Kubera himself is not a particularly attractive fellow, and indeed his name may mean "ill-shaped one." He is a white dwarf (!) with three legs, eight teeth, and a single, yellow eye (on the left). He has a large stomach.

The opinion of some historians is that these beings represent gods from the older, non-Aryan, peoples who inhabited India before the Aryan invasions, and who were "coopted" into the Hindu pantheon much the same way Buddhism later "coopted" existing Hindu and Tibetan Bön deities, and as Christianity did with Greek, Roman, and Celtic gods. Pulastya himself, father of Kubera, is deliberately identified as being non-Aryan, so this would bear the theory out.

We have spent so much time with Kubera and his yakshas because of another fascinating item: these beings—the Anti-Gods or Asuras—are experts in magic, and dwell both below the earth and in heaven. Indeed, the domain of the North—the polar quadrant of both the earth and the heavens—belongs to Kubera. It is believed that the word "asura" may be derived from the word "sura," which means "to shine," as does the word "dev," from which we get the words diva, divinity, and devil. This association with beings who dwell *both* below the earth (in an "Underworld") and in heaven, and whose generic name means "to shine," may possibly point to a stellar origin of these beings in a myth that is now lost.

But another theory on the word "asura" has it that it is not a Sanskrit word at all, but has other—so far unknowable—roots. May these roots possibly lie in Mesopotamia, land of Sumer and of the *Necronomicon*? May the root of "asura" be "Assur"?

Assur or Asshur is the ancient name for Assyria, a land to the north of Sumeria that was settled by people from Babylon. At the time of its greatest power, however, it comprised lands and civilizations as diverse as Babylonia, Egypt, Syria, and Persia. Around 612 B.C. their empire was destroyed by the rebellion of the Babylonians and the Medes (Persians).

Asshur, itself, was the name of their chief Deity and their principle city, which was later known as Nineveh. The leader of the Assyrians was also the chief priest of the Asshur cult, in distinction to the Sumerians and Babylonians, who kept the roles of priest and king separate. This identity of priestly and kingly functions in one person, however, was very similar to that of the Egyptian pharaohs whose land was invaded and controlled by the Assyrians and made part of the empire.

Other Assyrian deities included Marduk—borrowed from the Babylonians, who borrowed him from the Sumerians and Ishtar, the Babylonian name for the Goddess Inanna, and who later became a "demon" to medieval Christianity in the name of Ashtaroth. One of the most famous of Assyrian leaders was Ashurbanipal, who led that nation at the time of its greatest power on earth—shortly before its total destruction at the hands of the Babylonians.

This famous Biblical-era kingdom was noted for its excessive cruelty and warlike personality. Although the city of Asshur/Nineveh was renowned for its beauty and extravagance, the kingdom was not loved or admired. There are many missing pieces in the history of Asshur, hundreds of years for which there is no history, no record of what went on in the realm or what conquests took place or what intercourse there was with the other kingdoms of the world. But their idols of fabulous winged beasts and of kings flying in strange, solar, winged vehicles were well-known throughout the Middle East. They have been the source of much speculation and controversy, some—in the Von Dänniken tradition—even going so far as to propose that the Assyrians had somehow known of manned flight or even space travel.

Could the god Asshur have been the prototype for Kubera, Lord of the Yakshas, one of the clans of the Asuras? Could the enormous wealth of the Assyrian kingdom—plundered, for the most part, from the wealth of other lands—have given rise to the Asuras as being "Lords of Wealth"? The Tibetan *Book of the Dead* advises "Quarreling and warfare are the chief passions of a being born as an *asura*,"[16] an apt description of the way the Assyrian empire was perceived. Was there, at some time, communication between India and Assyria that could have led to either the Assyrians borrowing the concept of "ashura" from the Indians, or vice versa? We shall probably never know, but the consternation of philologists over the derivation of the non-Aryan word "asura" is what gave rise to these musings, for the Assyrians were, of course, Semitic-speaking people and therefore non-Aryan. (It may also help to remember, in this place, that the Sumerians were not members of the local Semitic tribes at all.) There was, of course, an Egyptian deity called "Shu," which means "bright" in ancient Egyptian as well, and the name for Osiris in Egyptian is "Asar." This name crops up in the *Necronomicon* and in the Sumerian names for Marduk as "Asaru," and appears in the names "Asarualim" and "Asarualimnunna." As the Sumerian civilization predates the Egyptian, one cannot help but wonder if the original Osiris was nothing more than a Sumerian import, and if both (or either one) were imported into India, where they were worshipped before the Aryan invasions transformed them into demons instead of gods. The linguistic correspondence of "shu" to "bright" among Egyptian, Sumerian, and Indian (not to mention Babylonian) sources in the names for a great god (Asar, Asaru, Shu, Asshur, Osiris, Ashura) leads us to believe that there was once an Ur-language and a correspondingly Ur-religion.

The Thigh of Set

As mentioned in the beginning of this chapter, the Big Dipper was known in ancient Egypt as the Thigh of Set. This wounding of Set is probably a sexual metaphor the same way the mysterious wound in the thigh of the Arthurian figure Amfortas is a metaphor: they are both euphemisms for castration. For instance, in one myth Osiris is depicted as having been wounded in the thigh, like the Fisher King of the Grail legend. This is viewed by the experts as a castration motif.[17] Moreover, another name for the Thigh of Set constellation is the Bull's Thigh. Set was depicted in this instance as a Bull tethered to a stake (the pole star) and hence walking in an endless circle around it. The bull as symbol of awesome strength and virility in ancient Egypt is well known. When the bull is "tethered" it has become, in a sense, castrated. Its power, its inherent wildness, has been tamed. The Egyptians referred to this constellation specifically as the "thigh" of the bull, and not the bull itself, even though the paintings show the entire bull. The implication, of course, is that the "thigh of the Bull" and the "thigh of Set" refer to the phallic—and hence the creative, libidinous powers—of these beings.

The word for "bull" in the ancient Egyptian language is *ka*. An honorific title for the Pharaoh is *Ka-mutef*, which means "the bull of his mother." This refers to the procreative aspect of the divine marriage between the pharaoh and his queen, who give birth to their successors. It is also a patently obvious reference to a kind of divine incest between mother and son, a ritual neutralization of the Oedipal complex that is at the core of so many initiation ceremonies.

The complex of ideas surrounding the Bull symbol, however, includes representations of Osiris in his aspect of a virile, potent deity and all the attendant myths dealing with his mystical impregnation of Isis after his death at the hands of Set. Set is also a bull, the Bull of the Heavens, the Bull who is slaughtered, sacrificed, so that the deceased can "open his mouth": become conscious once more.

If the ancients loved punning as much as writers like Temple insist, then in the case of the Bull mythos we have an excellent, and pertinent, example. For the word for "bull" in Egytian is *ka*, as we have mentioned, and the word for the spiritual "double" of the Egyptian religion is also pronounced *ka*. Although the hieroglyphics for each of these words are entirely different, the pronounciation is the same. Was this the result of a deliberate attempt to unite the two ideas? We may never know. The antiquity of the language and the religion that accompanies it forbids any definite answer. One thing is for certain, however, and that is the identical sounding terms would have provided the Egyptian priests with a rich source of ritual allegory that we would be amiss to ignore. That the Bull and the Spiritual Double—the *ka* that is both the essence and the occult power of an individual, whether peasant or king or god—are somehow reflections or manifestations of each other, is an idea that cannot be readily denied. The word for occult power in Egyptian is rendered as *Heka*, the *ka* syllable identical in hieroglyphic form to *ka* the Double. It is a pair of upright, raised arms, thusly:

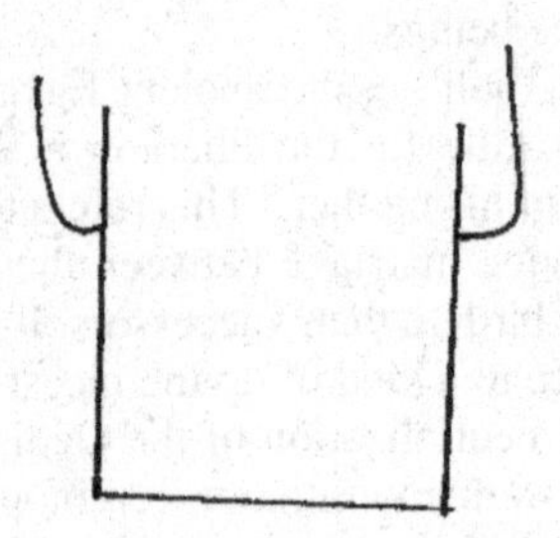

These upright arms formed the basis of some speculation in the edition of *Goetia: The Lesser Key of Solomon* prepared by Aleister Crowley. The seals numbered 162 to 174 inclusive in that book all show some version of a person raising their hands into the air. Crowley says this is "the evident desire to represent hieroglyphically a person raising his or her hands in adoration."[18] It would certainly seem that the raised hands owe more to the classical representation of the Egyptian *ka* than to a simple act of adoration, for *ka* is at the heart of all Egyptian magic. *Heka* shows this same symbol, accompanied by a hieroglyphic depicting a cord twisted three times, thusly:

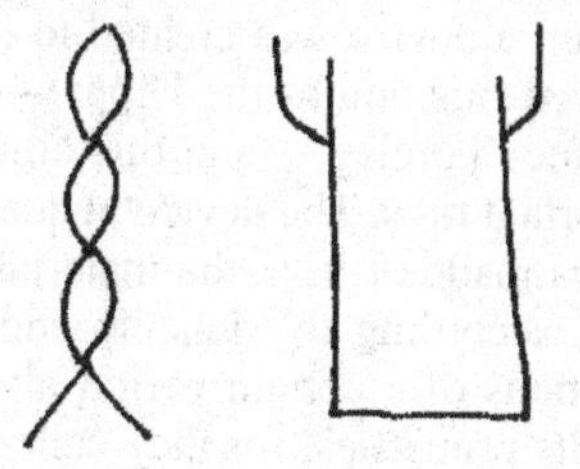

Would *heka*, magic, then have the meaning for the ancient Egyptians of a binding up of the *ka*, or spiritual double, of a person or a god? And is not this power somehow linked to the sexual power represented by the bull, the other *ka*? If so, the image of a tethered bull—as in the Bull's Thigh constellation—suddenly takes on new meaning as a graphic depiction of a supreme magical act, and the performance of magic becomes nothing less than a "tethering" of the power of Set, the Celestial Bull.

(For those readers familiar with the whole literature of Thelema and of Crowley in general, an item of some passing interest: you will remember the story of how Crowley was in the Cairo Museum in 1904 and became directed to a stele numbered 666 in the catalogue, the Stele of Revealing. During my researches for this book, I came across another such bizarre coincidence in my study of the Bull, for in Maspero's *Dawn of Civilization*, published in 1922, there is an illustration on page 119 entitled "The Sacred Bull, Hapis or Mnevis." It is intended to show by which sacred marks an animal might be recognized as the incarnation of a god. The number in the upper-left-hand corner of the illustration—taken from a work entitled *Notice des principaux monuments* by one Mariette, published in 1876 and appearing on page 222—is none other than the Mark of the Beast itself: 666. *Quod erat demonstrandum*!)

But the myth went further than a handful of stories designed to explain certain aspects of creation. In the case of the Thigh of Set, a device was created to imitate the shape and the perceived function of the Dipper—known in Egypt as *Meskhetiu*, the "Foreleg"—and this object was essential in several important rites. The device in question is the adze.

The adze was made of iron, the material of the "bone of Typhon" (Set), according to Manetho and Plutarch. However, the Egyptians of a certain period also were aware of lodestone and its properties, for they called it the "Bone of Horus." Lodestone, of course, is an ore of iron known to modern chemistry as Fe_3O_4, or magnetite. Although not all pieces of this ore have the ability to attract iron, those that do—called "natural magnets"—must have exerted a powerful fascination for the ancient Egyptians. There is no evidence to suggest that the pair of adzes—called, sometimes, *Neterti*: "the two divine adzes"—were made of lodestone. Yet, they were used in pairs, one for Horus and the South and one for Set and the North.[19] These directions do not refer to the terrestrial south and north, for in that case Set was the ruler of Upper Egypt, or the southern portion of that country. Instead, they refer to their celestial counterparts, with Set ruling the heavenly "north," the place of death and the Underworld, with Horus ruling the quarter of the heavens where the sun is highest, the south. (It is tempting to believe that at least one of these adzes was made of lodestone or contained a lodestone component, but I am unaware of any hard evidence to support this. Plutarch flourished around the time of Christ, so his Egyptians are much later than the earlier Dynasties—First to Fifth, roughly—with which we are most concerned.)

Hence, the divine adzes referred to a divine polarity, and were made in the shape of the Dipper and referred to by the Dipper's Egyptian name of "Bull's Thigh" or "Foreleg." As the Dipper also points to the celestial pole, the adze (which is a kind of small chisel, with the blade perpendicular to the handle) was fashioned in the shape of the Dipper.

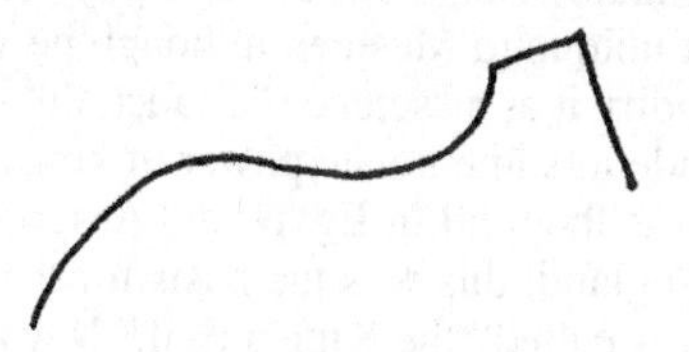

The symbol for the adze is a component part of many words of interest, such as *time*, *return*, *flood, hunters*, and *crookedness*. For now, it is important to mention a rite in which power was passed from the gods to the pharaohs, the Rite of the Divine Service. This power was referred to as *sa ankh*, the last syllable being that of the famous "eternal life" hieroglyphic, the ankh. The means of transferring this power from the god (Osiris, Ra, Horus, or Ptah) to the king was known as *setep sa*, a term that indicates "magical passes." The king would kneel before the God and the God would perform these magical passes, *setep sa*, behind him, running his hands from the nape of the king's neck down to his lower vertebrae, a path suggestive of the Hindu *sushumna* pillar analogous to the biological spinal column. The hieroglyphic for *setep sa* includes the glyph for *adze*.[20] The second part of this rite involved the king sitting on the lap of the goddess and drinking milk from her breasts.

The author submits that these "magical passes" are the ancient form of what have been called "mesmeric" passes in recent history, and which were described as a means of harnessing the body's natural "magnetism." This latter theory, developed by the Austrian medical man Franz Anton Mesmer (1734-1815), was called *animal magnetism* by him and was later better known as the practice that bears his name today: Mesmerism. Considered by some to be the origin of modern hypnosis, it was believed to be a function of a kind of electrical "sympathy" said to exist between living organisms, which could be used to influence their health by means of "mesmeric" baths and the strange gestures—similar to the laying on of hands—of the mesmeric physicians. This concept was not unique to Mesmer, although he was the first to attempt to codify it as a "science." In fact, this ability was believed to reside in some innate power of kings, who ruled by divine right (as they did in Egypt and Assyria, among other places). In England, this was the basis for the cure of scrofula, which was called "the King's Evil." It was believed that the King (and not the Archbishop, mind you) could cure this disease by the laying on of hands. Every monarch from Edward the Confessor to Queen Anne was supposed to possess this occult power. The French kings also claimed this ability, from the time of Clovis in the fifth century, A.D. Clovis was the king who consolidated French and German territories that had formerly been part of the Gaulish lands of the Roman Empire, and who converted to Christianity and then forced the rest of his subjects to follow him in his conversion. In France, as late as the seventeenth century, thousands of faithful subjects were brought before the king, who touched them, uttering these words: "The King touches you, God heals you," thus emphasizing that this kingly power was transferred along royal bloodlines straight from God.

In England, of course, this divine kingship began with the ur-monarch, the famous King Arthur, whom, as we have seen, may well have been nothing less than a mythic representation of that all-important celestial diagram, the Great Bear.

It was this constellation that the pharaoh used in setting out the orientation of the new temple of Denderah in the ceremony known as the "Stretching of the Cord." This was the ritual sacred to the goddess Seshat, the wife of Thoth, who is pictured with the seven-rayed star on an upright rod as her crown, beneath the inverted crescent of what might be the vault of heaven, or the horns of the bull from which the Bull's Thigh constellation gets its name. It is the author's contention that the seven-rayed star of Seshat's crown refers to the seven-starred constellation of the Dipper, and that the rod represents the adze in a different form. It is also our contention that the oddly inverted horns represent the fact that the bull is *upside-down* during the important ceremonies: i.e., "hanging from its tail in the sky."

The Opening of the Mouth

Our story might well have ended there were it not for the equally important ritual of the "Opening of the Mouth," a crucial element in the mortuary ritual preserved in the Book of the Dead. The Opening of the Mouth is the final moment of consecration in the lengthy funerary rites in which the essence—the *ka* or double—of the deceased is sent back into the upright, mummified corpse (or, in later times, a statue of the deceased) by touching its mouth with the adzes, signaling the instant of rebirth in the Other World. (If we care to revert to Egyptian style punning, we might say that the ceremony of the Opening of the Mouth, by using the Bull-symbol adze to send the Bull-like *ka* back into the otherwise helpless mummy, is a kind of cosmic "bullying.")

The whole point of the funeral rites of ancient Egypt was to ensure that the *ka* of the deceased would be reborn as was Osiris and would dwell among the stars in heaven. The moment this rebirth takes place is in the Opening of the Mouth ritual, and it involves the slaughter of a bull, the cutting out of its heart and its foreleg, and the anointing of the lips and eyes of the deceased (or the deceased's statue) with the blood of the sacrificed animal, among other ingredients. As this ritual is a key to understanding the thesis of this book, we will look at it in some detail.

The mummy of the deceased—or, as we have said, in later times a statue of the deceased—is set upright on a mound of sand. The mound itself is a potent Egyptian symbol, and may refer to a pre-Dynastic burial custom. Osiris is sometimes depicted as a mound of earth with its concomitant imagery of potency, of potential growth, and of a treasure buried within. In the rites we are describing, however, this mound is a small one, for the rite takes place in the deepest part of the burial chamber.

There are several priests of various ranks in attendance who perform specific aspects of the ritual. These may be examined in detail in the available literature. The *Kher Heb* and the *Sem* priests are the central players. The Kher Heb is the "priest who reads the scroll," i.e., the high priest and master of magical ceremonies. The Sem priest is a kind of assistant who performs the acts that the Kher Heb describes. A bull is slaughtered. Sometimes more than one. A bull is not a small animal. The important part of the bull for this sacrifice is the foreleg. The foreleg in this instance is a large extremity, perhaps three-quarters of the height of an average Egyptian or more, to judge by the wall paintings that depict the various actions of the ritual.

The Thigh of the Bull used in the Egyptian Ritual of the Opening of the Mouth. From *The Book of the Opening of the Mouth* by E.A.W. Budge.

The Sem priest lifts up the leg and approaches the mummy or statue. In some texts the Sem priest is told to touch the mouth and eyes of the deceased with the bloody foreleg, thereby smearing some of the blood on the face of the statue. In an interesting part of the accompanying chant are the words:

> Hail, Osiris! I have opened for thee thy mouth with the Leg, the Eye of Horus.[21]

So, the Foreleg of the Bull, the Bull's Thigh constellation, is also the Eye of Horus. This identity is reinforced by the next stage of the ceremony, in which the Sem priests lifts the two adzes, one callled Seb Ur, or the "Great Star," and the other simply An, a reference (according to Budge) to a star in the Great Bear constellation. One of these instruments represents Set, but the other represents Horus.

The Ur-Hekau Instrument used in the Egyptian Ritual of the Opening of the Mouth. From *The Book of the Opening of the Mouth* by E.A.W. Budge.

The Sem priest takes the first instrument and slits open the mouth and eyes of the mummy (or mimes this with the statue), and recites a prayer telling the deceased that he has opened his two eyes and his mouth, ending with:

> Hail, Osiris! I have opened for thee thy mouth with the instrument of Anpu [Anubis].
>
> I have opened for thee thy mouth with the divine instrument, with the MESKHA [i.e., Leg, or Thigh] of iron wherewith the mouth of the gods was opened.[22]

Budge goes on to tell us that the iron adze, referred to in the prayer as belonging to Anubis—Lord of the Dead—is to be considered equivalent to Set, "for the instrument which he used was made of iron, and iron is a Typhonic metal."[23]

He refers here to the same statement mentioned above by Plutarch, in the latter's *De Iside*.

Then the Sem priest picks up the adze sacred to Horus and says:

> Horus shall open the mouth of the Osiris, even as he opened the mouth of his father.
>
> As he opened the mouth of the god Osiris, so shall he open the mouth of my father with the iron which cometh forth from Set, with the MESKHA [i.e., Leg or Thigh] instrument of iron, wherewith he opened the mouth of the gods, shall the mouth of the Osiris be opened.[24]

Again, the instrument is of iron "which cometh forth from Set" and is being used by Horus to open the mouth of "the Osiris": that is, the deceased in his aspect of a reborn soul, or *ka*. The iron "cometh forth from Set" may be a play on several concepts, for the Bull's Thigh is a constellation and the actual foreleg of the slaughtered bull, as well as a phallic symbol. Symbolically, therefore, the freshly butchered bull *and* the Northern Dipper *and* the ritual adze are *all* Set, and iron "cometh forth" from Set. This may refer to the "directed panspermia" of Dr. Crick, as well as to the creation of the human race by an essence that "came forth" from Marduk in the Sumerian legends. As the incantation shows, the word for Bull's Leg or Thigh is *Meskha*. The word for the *constellation* of the Bull's Thigh is *meskhetiu*. In hieroglyphics the word for Bull's Thigh constellation and Bull's Thigh adze are composed of the same elements, save that in the case of the bull an actual bull is suffixed to the word, and in the case of the adze the adze symbol is suffixed. These pictorial "suffixes" seem to have no phonetic relevance to the word, and are only added as clarifiers.

A further point of information regarding a strange locution above: the Sem priest is speaking to the deceased as if he were the son of the deceased. This is a mime of an ancient myth that has somehow been lost, in which Horus (the son) opens the mouth of Osiris, his father. In fact, the origins of this entire ceremony are buried deep within the pre-Dynastic, preliterate period of Egyptian history, or perhaps somewhere at the moment when the Egyptians suddenly *became* literate. We don't really know what this aspect of the myth actually means, based on the textual evidence available. We can only conjecture at this point, based on the symbols themselves, which are quite powerful and amenable to interpretation once other cultural inputs are received, as we shall see.

To continue with the ritual, the Sem priest now employs another device, an ebony rod called the *Ur-Hekau*. You may remember that *Heka* is the word the Egyptians used to denote magic power. Adding the "u" suffix changes the word from singular to plural. The prefix "Ur" means "the Great." Budge translates *Ur-Hekau* as "the great one of words of power."[25] This instrument is "a wooden staff made in the form of a small serpent, with the head of a ram, which was surmounted by an uraeus."[26] The uraeus, of course, is the small serpent, perhaps an asp, that we see on the headdresses of the kings and queens of Egypt. Hence, this most potent *occult* symbol has the body of a serpent, the head of a ram, and a serpent crown.

It sounds suspiciously like the caduceus of Hermes, the staff with two serpents coiled around it that is surmounted by a winged globe. Exchange the pair of wings for a pair of horns and you have a reasonable facsimile of the caduceus. We know the caduceus in the West only as the symbol for medicine, although, as the staff of Hermes (the Greek Thoth), it also represents occult knowledge and power. In the "Opening of the Mouth" ceremony, we are pleased to discover that this very staff is used in a *healing* ritual. After the two adzes have been employed to open the eyes and mouth of the deceased, the *Ur-Hekau* is used to heal the wounds made by the slitting of the flesh and to stanch the flow of blood.

How much more detail do we need in order to draw the obvious inference, that the *Ur-Hekau* is the caduceus, the symbol of the magick wand *par excellence*? The Author has great familiarity with this type of wand, for in his ecclesiastical period the caduceus was the staff of his office. It is a common *crosier*, or bishop's staff, among the Eastern Orthodox churches where the tradition has it that the wand encoiled by two serpents is the symbol of the rod of Aaron, which confounded the Egyptian priests and was instrumental in getting the Pharaoh to permit the Jews to leave their captivity in Egypt. Naturally, the basis of the crosier's design is far older than that, as we can see from our example.

This identical symbol can be found in the hand of Cernunnos, on the famous Gundestrup Cauldron that was recovered from a bog in Denmark. There is tremendous controversy at this time as to the provenance of this device, with certain modern archaeologists finding evidence of a shamanic cult that extended from India to the Celtic lands, through the Middle East.[27] Certainly, a ram-headed serpent staff is a bizarre enough instrument that it would not have been independently concocted by various disparate tribes of itinerate craftsmen through the millennia!

The chant employed in this part of the ritual is well worth repeating for its wealth of information:

> The goddess Nut hath lifted up for thee thy head.
>
> The god Horus hath taken possession of the Urerit Crown and his words of power [i.e., *Hekau*, ed.].
>
> Behold, Set hath taken possession of the Urerit Crown and his words of power.
>
> Behold, the goddess Nut hath appeared with thy head.
>
> All the gods bring [words of power], they recite them for thee, they make thee to live by them, thou becomest a lord of twofold might, thou makest the

> passes which give thee the fluid of life, and their fluid of life is about the Osiris.
>
> Thou art protected and thou shalt not die.
>
> Thou shalt make thy transformations among the KAU [or Doubles] of all the gods.
>
> Thou shalt rise up as a king of the South.
>
> Thou shalt rise up as a king of the North.
>
> Thou art endowed with strength like all the gods and their KAU.
>
> And behold, this statue of the Osiris is SHU, the son of Temu, and as he liveth even so shalt thou live. . . . Thou hast made the passes which place life around about the image of the Osiris.[28]

This chant is interesting because it is a recap of all that we have been speaking of, neatly wrapped in a single package. We begin with references to the goddess Nut (Nuit) and the god Horus, as well as Set, all of whom partake in this resurrection of the Osiris (the deceased is referred to as "an Osiris" and, in fact, the word Osiris itself may be found to refer more to a *function* than a static *being*). Before the invocation is over, even Shu will be remembered as well. Shu is the Sky that holds up the Heavens: Nuit. Horus we know as the avenging son of Osiris, and Set is the being who killed Osiris and from the power of whose Foreleg Osiris will be reborn. The Foreleg itself is referred to as the "Eye of Horus": probably the most potent religiomagical symbol the Egyptians ever possessed, and a description of which would take up a volume equal in size to this one.

The Osiris is referred to as a "lord of twofold might," which is a reference to the deceased as a newborn "King of the South" and "King of the North," i.e., the twin celestial poles represented by the twin adzes that are made of the iron of Set—the Typhonian metal sacred to the sexual, chaotic, Dionysian force of nature. Twice in the above invocation reference is made to the "passes" that the God is believed to make over the body of the deceased to restore it to life. We remember these passes from the daily ceremony in the temple of the Pharaoh, and they are identical.

Let's see what Budge has to say about them:

> The use of the UR-HEKAU instrument produced a wonderful effect, for, through the four passes which the Sem priest made with it, some of the vital power, *sa ankh* . . . of the gods was transferred to the statue. Moreover, knowledge of the words of power which were known to Horus, and Set, and the gods, was given to it, and by the use of them the deceased became king of the South and North, and lord of twofold strength, and the vital power of the gods was round about him on all sides.
>
> The vital power . . . could be transferred from one god to another, or to a king, or from a living being to a statue, by the touch, or by "making passes" over the neck and down the spine of the person, or statue, or by an embrace. In this case the vital power, or "fluid of life," which the gods Horus, Set, Thoth, and Sep had derived from Ra, was transferred from them to the statue by the SEM priest, through the agency of the instrument UR-HEKAU. The ram's head on this instrument symbolized Khnemu, one of the oldest of Egyptian gods, whom religious tradition declared to be the "builder of gods and men, and the Father who was in the beginning, the creator of heaven, earth, the Other World [Tuat], the waters and the mountains, who set up heaven on its four pillars." Now Khnemu contained the souls of Ra, Shu, Seb, and Osiris, i.e., the souls of the greatest of the gods, and the power of all of these was transmitted to the statue by the passes made with the UR-HEKAU . . .[29]

I have quoted the above at some length for it touches directly on our subject. The *sa ankh* power is passed from the primordial creator Gods down through the UR-HEKAU, in tandem with the twin adzes, and transmitted by means of magical passes over the body, or statue, of the deceased. Moreover, the picture that accompanies this stage of the rite shows the UR-HEKAU in profile, and it is nothing more in shape than a somewhat larger version of the adze itself.[30] It is said several times in the liturgy we have just studied that the iron adze "opened the mouth of the gods." That is, in some previous time long ago, this ceremony was preceded by an actual or mythical *event*: the iron instrument in the shape of the Dipper was used to "open the mouth of the gods," to give life to the gods. A review of Egyptian mythology and ritual shows us that sometime after Khnemu, who fashioned the gods Ra (Sun), Shu (Sky), Seb (Star), and Osiris (the Risen God), their mouths were "opened" by this magical instrument. And, by applying it to the body of the *deceased*, this pact is commemorated with the "opening of the mouth" of some mere mortal who may now take his or her place among the company of the gods in the heavens, i.e., become one with the stars, which are the gods *in essence*. Hence, the *ka* returns to its *place of origin*, symbolized by the Iron Adze, the Great Bear constellation. The *ka* returns to the creator, follows a line of succession or heredity down the generations and through to the Other World.

It is worth noting at this point that, in the *oldest* legends coming out of Egypt regarding the Underworld (Tuat), the Underworld is composed of *seven gates*. It is only in later epochs that the Underworld is viewed as having as many as twelve or more gates.

Here, in the ceremony of the Opening of the Mouth—a repository of some of the oldest Egyptian practices and beliefs concerning the Other World—we have found a reverence for the adze as an instrument of astounding power: nothing less than the ability to reawaken the soul of the deceased, to give rebirth in the Other World. We have found that this instrument is made of iron, is sacred to *Set*, is consciously made in the shape of the Big Dipper, and is called by the same name as they called the Dipper: the Bull's Thigh constellation. Indeed, they went so far as to slaughter a bull during the ceremony so that they could wipe the blood from its severed Thigh on the mouth and eyes of the mummy. The deceased is then proclaimed a lord of South and North; two adzes are used in the opening, one for the South and one for the North; magical passes are made along the spinal column of the deceased to transfer the sacred power from the gods and to awaken and protect the deceased, and that these passes are made with an instrument that so closely resembles the caduceus of Hermes in form and function as to be virtually identical.

We have also seen that the iron adze, the MESKHET, is also crucial to the alignment of the temple of Denderah and probably to other sacred spaces as well, as it is referred to so constantly, and that the magical passes made over the body of the deceased are an echo of the passes made by the gods over the body of the Pharaoh for the transfer of *sa ankh*, and that these passes are referred to as *setep sa* . . . a term that includes the hieroglyphic for "adze." It is worthwhile to note that one of the four Egyptian goddesses of birth was MESKHENIT, who sported a headdress in the shape of a "cow's vulva" according to Budge, but which shape is suspiciously similar—virtually identical—to miniature examples of the *Ur-Hekau*. This is mentioned to show the Egyptian emphasis on the birth/death/rebirth/sacred bull complex.

Prior to this "awakening" or "opening," the deceased has passed through seven gates in the Underworld and undergone various trials. After this ceremony, the deceased will leave the tunnel of the Underworld and reappear—not in *our* world—but in the heavens as a Star, an *Asar*, an Osiris. The Underworld, therefore, seems to be entered by the deceased from *our* side upon death and exited on *another* side upon rebirth, like a wormhole in space. The after-death experiences familiar to modern readers of tabloid journalism seem to echo this concept of a tunnel with light at the end of it for the soul, or *ka*, of the deceased individual to travel.

The primal nature of the adze is reflected in other, more overtly magical, practices that have been recorded elsewhere[31] and which include a spell against serpents in which an instrument, "probably an adze," is used to mimic an ancient battle between Atum and the Serpent Neheb-Kau, the primeval serpent god also known as Amun "the Invisible One." In this case the adze is said to be "the claw of Atum" that defeated the Serpent sometime before humanity was born.

The magical power itself—the *Heka*—is said to have been brought here from "a distant, magical source." This source is the "isle of Fire" that exists "beyond the limits of the world," where the gods are born and to where they retire. A place of origin, and of return. The famous Phoenix of the Egyptians is said to be a messenger from the Isle of Fire, a circumstance that led to some confusion on the part of later historians when they described the Phoenix as a bird that died consumed in its own fire and was reborn again from its ashes.

In a related myth, Osiris is wounded "in the thigh" and his prone body partially conceals a sacred scroll. When Osiris is moved and the scroll revealed from under his thigh, it is found to contain the secrets "whereby the mouths of the gods are opened."[32] Again we have the motif of a god wounded "in the thigh" and its connection to a means of Opening the Mouth—a quickening, a calling back to life—of the gods.

What is this device? How does it "open the mouth" of the gods and the pharaohs? Why would a piece of iron appear to have the power to reanimate dead matter, at least symbolically? What is the connection between the Thigh of Set and the *ka* or spiritual double of a human being? How are the souls of the dead connected to the Big Dipper, that group of stars known in India as the Seven Rishis, in China as the Celestial Gate, the place where humanity has its origins?

And why does the "magic wand" of Western ceremonial magick contain pieces of lodestone at either end or a rod of magnetized iron throughout its length *to this day*?

To understand how all these seemingly disparate elements are related, we need to find out more about how the Egyptians described their Underworld, and the means they employed for breaching its walls. We need to find the Egyptian "gate."

The Underworld and the Seven

To the Egyptians, the Underworld—a word transliterated variously as DWAT or TUAT or DAT—lay under the earth, as in most early religions. However, they simultaneously perceived the Underworld as existing in the sky, beyond the planets and the zodiacal signs, just as the sages of ancient India believed the Seven Rishis to exist both below and above the earth. Our term "underworld" may be, therefore, misleading. Instead, a certain kind of primordial plane or level may be implied by the concept of DAT.

The Egyptian Underworld realm was populated by seven faceless, formless beings who nonetheless are depicted as quasihuman and wearing horns. This may be an analogue to the Seven Rishis of India who, as the progenitors of humanity, may be regarded as denizens of a suitably primordial plane themselves. In Sumeria, also, the Underworld was the habitation of the dread Annunaki, *seven* lords who are similarly faceless in that they are given no independent names or personalities, but instead symbolize the primeval forces that shaped the present universe.

Seven Lords in Egypt, Sumer, and India who are responsible for creation in its earliest stages. Seven Lords who dwell in the "Underworld": an Underworld understood by both the Egyptians and the Sumerians as existing simultaneously under the World and in Space beyond the planets. To the

Sumerians, specifically, the Seven Annunaki of the Underworld were synonymous with the Seven, equally faceless, Igigi of the realm of the "Fixed Stars."

This concept of "Seven" in its relationship to the Underworld is very nearly universal. There were seven caves in the Underworld of the ancient Aztec and Mayan cultures (exemplified by the seven-pronged catacomb beneath the center of the Mayan Temple of the Sun at Teohuahtitlan, and the seven caves that were the mythical origin of the Aztec race), just as there were seven caves or gates in the Underworld of the Sumerians and the Egyptians. Simultaneously, this "Sevenness" was echoed in the fixed stars, in the realm beyond the planets: specifically, in the constellation of the Big Dipper or Celestial Gate. This sevenness found its expression in the rituals and beliefs of the Daoists of China, as well. And in Tibet, the Bardo state (the "Between the Two Worlds" State after death and before rebirth) lasts for a symbolic forty-nine days, or seven times seven days: a representation in *time* of what the Sumerians, Egyptians, Chinese, and Mexicans represented in *space*.

And, as far away as the British Isles, the Celtic hero Cu Chulain travels to the Land of the Shade—the Celtic Underworld—and finds a fortress surrounded by seven walls. Naturally, this Underworld realm has its share of flying serpents and monsters. From Asia through India and Tibet to the Middle East, Africa, and Britain, and across the oceans to Central America, the Underworld is consistently a place—or a time—divided into seven parts. This means that *all* of our ancient civilizations recognized an insistent theme of Sevenness when it came to depicting the structure of the Underworld. Where did this universal concept come from?

The pyramids and ziggurats of the Egyptians, Mexicans, and Sumerians were ladders in stone that lead to the sky. The earliest were built in seven great steps, one for each of the Zones. This finds its resonance in the Seven Chakras of Tantric yoga, which is a ladder of sorts for the Serpent Kundalini (dweller in the Underworld of the human body) to ascend in its reach for heaven and divine illumination. The ziggurats of Ur are an external manifestation of what takes place in the body during the successful pursuit of what in India is known as Kundalini yoga. The chakras of Kundalini yoga represent internally the external voyage of humanity up the ladder of lights to its celestial source. Quite possibly, one did not *predate* the other, in the sense that the chakras were somehow *invented* in imitation of the seven stepped pyramids of Ur, nor were the ziggurats invented to give external form to a reality perceived to exist within the body.

In the Egyptian "Book of What Is to be Found in the Underworld"—a series of paintings and hieroglyphic scriptures found painted on the walls of the great tombs—there is a depiction of the last phases of transformation a soul goes through in the Underworld. It has baffled scholars to this day, as no attempt is made to clarify these particular pictures with words. It shows a man sitting on a serpent, and the serpent making its way to the stars. This is the penultimate stage of the soul's transformation. In this, the figure is accompanied by another serpent, this time with four (human) legs and a pair of wings. No one seems to know what this means.

At the last stage, the soul finds itself at the end of the "tunnel" with light coming from outside, and it is said that the tunnel is the womb and that the soul is prepared for rebirth.

The contention of this book, as anyone can figure out at this point, is that these are multiple manifestations of a *single* event: an event that takes place *both* in the microcosmic world of our psyches, our bodies, *and* in the macrocosmic world of human origins and destinations. The map of the Underworld is written not only in the sacred scriptures and temples of the Sumerians, Egyptians, Indians, Tibetans, Chinese, and Mayans (not to mention Malekulans!), but also in ourselves. We are the map. We are also the travelers.

To those who say, "It's all in your mind," we answer, "That's exactly right." To those who say, "It's spacemen from another galaxy," we also answer, "That's exactly right." Both *are* right. To an occultist, the concept of UFOs and "directed panspermia" or something similar, and that of angelic and demonic forces evoked by ceremonial magick, and that of internal alchemy or yoga to raise one to unity with Godhead . . . these concepts are not mutually exclusive. Unfortunately, that's what gives occultists the reputation of being "fuzzy thinkers" at best or absolute lunatics at worst.

During the Watergate scandal, Deep Throat is said to have remarked to Woodward: "Follow the money." In our investigations into the closely guarded secret of a different Gate, we can advise: "Follow the numbers." The insistence of the ancient peoples in the number Seven as being of paramount importance in a discussion of death, outer space, the Celestial Gate, immortality, and the genesis of human origins, cannot be due to some kind of coincidence. There are, after all, many other numbers to choose from! Why not five, or six? Or some other prime number, like seventeen, for instance? Or why not ten? Ten fingers, ten toes. Ten—or even twenty—seems a likely choice as a symbol of completeness, of perfection. Why seven?

If the reader has been following the argument so far, it will doubtlessly become obvious that there are seven visible stars in the constellation of the Big Dipper. And it is to this constellation that the ancients keep returning whenever the discussion evolves into a study of human origins, of death, and of the defeat of death and the return to origins. The seven stars are a *fact,* not a philosophical invention. The seven stars were there *first*. Philosophy came later.

And in *all* the philosophies under discussion, one begins at the first of the seven Gates and works one's way through them all, in order. And when one reaches the last Gate, an experience takes place that *pivots* the individual around and sends him or her back through the seven gates—or caves, or worlds, or planes—into a new birth. And in *all* the philosophies under discussion, there is a Serpent or Monster dwelling in the Underworld that must be dealt with before victory can be claimed. Whether that Serpent is the Goddess Kundalini, or Queen Ereshkigal or the Egyptian Wer or Apophis or Typhon or the Serpent destroyed by Quetzalcoatl from which humanity was created, or the Serpent Tiamat destroyed by Marduk from which humanity was created . . . it is always an Underworld monster and fiend at the end of the Seven Gates, at the center of the labyrinth.

In the Aztec myth of Quetzalcoatl, the god dies and travels through the Underworld, only to finally reemerge on the back of a winged serpent. Indeed, his very name means "plumed serpent": a serpent with feathers. A serpent with wings.

In the Sumerian epic, the Goddess Inanna (later, Ishtar of the Babylonians) descends into the Underworld and, at each of the Seven Gates, must remove an article of clothing, until she arrives naked at the end of the journey. In Asiatic shamanic rites, the descent into the Underworld is accompanied by more brutal losses: one is torn apart by demons, dismembered, disemboweled, decapitated, and put back together after one's inner organs have been "washed." The self is torn apart and put back together again, at the end of which the self ascends up a pole or a tree to the stars.

This gradual destruction of the self as a prerequisite to magickal power and wisdom is a constant through all of these Underworld myths, when the Underworld journey is undertaken as a form of initiation. We find this image in the published work of the late Scottish psychiatrist, R. D. Laing, for whom a "nervous breakdown" could also be a "nervous breakthrough," and whose descriptions of the schizophrenic experience were laded with references to initiatory metaphors.[33] To Dr. Laing, a schizophrenic patient was a shaman in the act of becoming. Eliade and other authors on mythology would probably agree. Even Freud, the founder of psychoanalysis, likened his career to that of the archaeologist and miner: a digging through the strata of the mind to reach the lowest layers of the unconscious; a concept that involved both space *and* time. His colleague Breuer—with whom Freud published *Studies On Hysteria* in 1895—likened the mind to a building with dark, underground cellars, an image straight from Sumer, Egypt, and Mexico.

At each Gate, therefore, something is removed from the novice. In primitive, megalithic cultures such as the Malekulans mentioned above, this "something" is the common sacrifice: of a tusked boar, or a pig (the "death pig," as it was called), or even an illegitimate child at the age of puberty. (In the ancient Syrian culture, the Big Dipper was called "the wild boar.") Among other cultures, the purely *interior* form of psychic self-immolation and subsequent reintegration became the *exterior* form of sacrifice on a sometimes massive level. The bloody human offerings of the Aztecs comes readily to mind. In this case, the "sacrifices" were often captured enemies . . . or their own citizens chosen by lot.

In the case of the worship of Cybele and Attis, of Artemis and of Astarte (Ishtar/Inanna), the sacrifice required of the novice was nothing less than his castration. It seemed to go without saying that the male worshipper of a Great Goddess had to become a eunuch "for the sake of Heaven." One thinks of the implied castration motif in the story of Set, who is chained to a stake held by the Great Goddess Isis in her form of Ta-Urt, the Great One. One also thinks of Osiris, who was castrated by none other than Set himself but who nonetheless was able to magically impregnate Isis so that she might conceive Horus. To a male, castration might signify the greatest "dismemberment" of all, an ample metaphor for the destruction of the ego (or the id?) taken to a barbaric conclusion.

Yet, Death is the greatest sacrifice of all. It's the sacrifice required of an individual if he or she desires to enter the Underworld in the normal manner, i.e., after a full life and in old age. Dying in battle was another way to quickly enter Paradise, as the Valhalla of the Norsemen and the Garden of

Delights of the Muslims would seem to insist. However, if one wished to visit the Underworld without undergoing the otherwise necessary loss of life, then it was imperative to offer a substitute sacrifice or series of sacrifices while on the Journey. One had to experience death *psychically* before one could master the Underworld and control one's own spiritual destiny. What better late twentieth century description of psychic death than the loss of one's ego?

In the Tibetan scenario, one's accumulated *karma* acts in much the same way as the Malekulan accumulation of *ramats*. It enables one to pass easily through the Underworld stages. Also in the Tibetan scenario, the more one has practiced yoga and meditation and attained advanced levels of psychic awareness, the easier it is for that person to remain oriented in the after-death state and to control its outcome. In other words, it is possible to experience death while still alive. It is possible to enter the Underworld realm while still in possession of all of one's faculties. It is possible to pass through the Gate—for all practical intents and purposes, to *die*—and to *return* to tell the tale.

It is upon this concept of Death, and existence after Death, that the whole edifice of religion is built. It is the common fear of Death that fills churches on Sunday and fuels the ecclesiastical economy. Death and the Life Hereafter is the proper territory of a religion that cannot empower its people in *this* life. Religion takes life-after-death as a given: it is the only "place" where its God may be found. For an occultist, however, nothing is a given. The vast unknown realm of prebirth and after-death existence is a territory to be explored immediately with whatever tools are at one's disposal. There is no smug acceptance of an inevitable human fate or destiny. The occultist does not rely upon reincarnation or tomorrow's heaven to cushion the blows of existence today. As J. Northcote Parkinson observed: "Work expands to fill the time available for its completion."[34] This applies to the *Great* Work, too. If one feels one has limitless lifetimes in which to attain nirvana, one will take limitless lifetimes.

For an occultist, there are no such guarantees. There is not even the guarantee of an immortal spirit or soul: no integral part of the human entity that will survive death to experience heaven, hell, or reincarnation. For an occultist, there is no God-given assurance of individual immortality. For an occultist, there is only the Gate, to be opened now, while still alive in *this* lifetime, in *this* body. With or without the approval of the Church, the State, the People. With or without the recognition of Science. For all of these are but aggregates of human beings, and an individual's potential for immortality is too precious to be left in the hands of the organization men, of the establishment, of the Citadel.

CHAPTER 3: The Great Bear: Key to the Gates

An enormous amount of attention is paid by all astrologers and most occultists to the planets, the Sun and Moon, and the twelve constellations of the zodiac. What is little understood today is the tremendous importance formerly paid to the most familiar constellation in the northern hemisphere: the Great Bear, and its component asterism, the Northern (Big) Dipper.

While the planets, the luminaries, and the zodiac were used as a kind of cosmic clock, useful for telling macrocosmic "time" and for predicting future events, the Great Bear served another purpose entirely: it was the means whereby the initiate entered the supernatural realm. It was the Gate to the Other World. To the Egyptians, the Dipper (or the Bull's Thigh constellation, as it was known to them) ordered the universe, set the cardinal points in their proper directions, arranged the seasons. It was the heart of the cosmic clock, the very center of the macrocosm. It was the origin of life.

Perhaps nowhere else is this attitude more prevalent than in the writings of the ancient Chinese sages, the Daoist initiates who developed complex techniques for entering that Gate and for aligning themselves with its power. In order to better understand the importance of the Great Bear in the cosmological scheme of the *Necronomicon*, we must first examine these writings that predate the authorship of the *Necronomicon* by at least eight hundred years.

* * *

The Daoist (Taoist) sages understood there to be a horizontal axis running through the earth, and a vertical axis running from the earth to the sky. Like the Sumerians—whose civilization was older than the Chinese civilization—they believed in the cosmic polarity of heaven and earth. This polarity was extended to include the human body itself. Our own bodies—compounded of the physical and the mental aspects as we understand them—were called the "yin" bodies. We received them passively, at birth; they function automatically, without our conscious intervention. In a way, it is the slate on which our lives are written. The yin body is subject to the pressures of heredity, society, and the environment. It is the "Square Table" of earth. It is our identity in its "organic" state, as Freud might have put it.

The goal of Daoist meditative and magical practices, however, was the creation of a (inorganic?) "yang" body: a mystical, numinous body graced with immortality and magical powers. Many readers will recognize this body in the "astral" or "subtle" body of the Theosophists and the later Golden Dawn magicians; the Middle Pillar ritual, for an example, is an attempt to formulate and strengthen such a body, a "body of light." The difference between the ancient Daoist concept, however, and the Western, Golden Dawn version is dramatic: to the Daoists, there was no immortality without the conscious *creation* of such a "yang" body. In other words, the vast majority of human beings die forever in this worldview: no heaven, no hell. Nothing. No existence after death is possible without the creation of one's own particular "soul."

Of course, with the passage of time and the introduction of Buddhist concepts into China, the idea of reincarnation took hold and Buddhism and Daosim came to be practiced—at first, side by side, and then together, intermingled as virtually one faith. Eventually, the creation of the "yang" body became viewed as something other than the actual spirit (which could, after all, become reincarnated after death), and became a lot closer to what we in the West would refer to as an "astral body": an engine of occult power and knowledge.

Those who are familiar with the Qabalistic Cross, the Lesser Banishing Ritual of the Pentagram, the Ritual of the Hexagram, and the Middle Pillar exercises of the Golden Dawn—not to mention the Star Ruby and similar exercises of the A..A..—would come to a rather obvious conclusion if asked to define the common denominator in all these rituals. Reference is constantly made to a Star, shining above the magician, whose beneficent rays bathe the operator in light and power. This common thread of meditation on a metaphorical Star in the sky would be quite familiar to a Daoist magician operating in the first century of the common era. To the Daoist, this Star could only be found in the greatest constellation of them all, the Great Bear (or Northern Dipper), with its companion star, Polaris.

To the sect of Daoism known as Shangqing Daoism, the practice of meditation on the Northern Dipper is virtually identical in spirit to that of the various Star rituals mentioned above. As will be discussed in detail in the next chapter, one imagines the constellation as descending from the heavens to position itself before, behind, and around the magician in various patterns. The Northern Dipper was the Gateway to the Underworld, the field of creative transformation, the path of celestial travel. It was also a protective power, keeping the practitioner safe from all manner of demonic influences that were sure to become attracted to the practitioner's occult studies and rituals. It was invoked in the North, the place of Death, and was believed to be a mirror image of the lowest point on earth, the place deepest below the surface of the earth. This nadir was, therefore, the Underworld where the dead were buried. To invoke the Northern Dipper, then, was to open the Gate between the living and the dead *and the gods* . . . but to do so while still alive, to gain control over the forces of death: to gain nothing less than immortality.

To the Chinese, their magnetic compass needles were said

to point *south*, not north. South was the important direction, the place where the sun was at its strongest, the place of high noon. Naturally, compass needles point in both directions; it's simply a matter of perception and preference to say that a compass needle points *south* rather than north. But the diviner's compass of the Chinese sages and practitioners of *feng shui* (what we have erroneously called in the West "geomancers") uses a needle that is shaped in the form of a spoon, or ladle, to represent the shape of the Northern Ladle, the Dipper, just the way the Egyptians fashioned their magic adze. The handle of the ladle is the pointer, and it points south. That means that the bowl of the ladle points north.

Hence, to the ancient Chinese as well as to the necromancers of the *Necronomicon*, the sacred direction is when the "Bear hangs from its tail in the sky," for that is when the Dipper's handle points south and its bowl points north!

Further, this instrument was used almost exclusively by the *feng shui* magicians who were responsible for orienting the Chinese temples. It was only later that the principles of *feng shui* (Wind/Water, a method of determining the correct locations and orientations of buildings, etc.) became accessible to everyone, and it is still practiced extensively today all over Asia wherever there is a Chinese community. The importance of orienting a building so it conforms to a direction determined by the earth's magnetic field and the position of the Northern Ladle or Dipper—seven stars in the nighttime sky—was understood by both the ancient Egyptians of 2000 B.C. and the Chinese sages of 300 B.C. It was important—necessary—to form a physical link between this World and the Otherworld by using the stars. The ancients knew that there existed a relationship between the earth's magnetic field and poles and the positions of the stars. If care was taken to position the doors, walls, and altars of the temple properly, then a Gate would be opened in this World that permitted contact with the Other, and the ways of the Gods would be made known on earth and the power of the Gods could be harnassed. The fact that *both* the ancient

Egyptians and the ancient Chinese fashioned a *tool*, out of *iron*, in the shape of the *Dipper*, means there is some common body of occult knowledge that has become all but lost in modern times.

The author submits that the Star rituals of the Golden Dawn and the A..A.. are but pale imitations of the more precise, more complex, invocations of Shangqing Daoism. This is not to denigrate the very great power and importance of the Western rituals. Rather, it is to show how these same rituals may be enhanced by understanding the lengths to which they can be taken by the careful occultist. Does not the term "astral body" refer to a body made of starry stuff? To the ancient Chinese, this term would have been immediately understood as the body created through invocation of the Great Gate, the Northern Dipper, the "body of light." The Invocation of this Gate becomes the initial rite of the *Necronomicon* and the starting place for our study of the occult system described in its pages. What the author describes is a complete magickal system that includes practices that can only be described as yogic, together with meditative and ritualistic methods for self-transformation on a scale heretofore unrecognized in the West.

The Three Gods of the Gates

Just as there are three Gods that must be invoked when attempting to enter the Gates of the *Necronomicon*—Anu, Enlil, and Enki—so there are three gods that preside over the Gate of the Northern Dipper in Shangqing Daoism: they are referred to as the Three-in-One, and are sometimes described as the Male, the Female, and the Great One; or the White Tiger, the Female One, and the Primordial King, respectively. Oddly, the name of the Third of these Three-in-One is *Ying'Er*, a name that sounds suspiciously close to the Sumerian *Enki*. He is described as the Ruler of the Gate of Life, and called the King of the Yellow Court in Chinese Daoism.

The Haitians also perceive a triple category of being that involves the *ti bon ange*, the *gros bon ange*, and the *n'ame*. To the Chinese, these may represent spirit, energy, and essence, respectively, and thereby become Caribbean equivalents for the Daoist Three-in-One.

During the course of the year, the Dipper changes position in the heavens, revolving around the Pole Star in a kind of cosmic dance. During the evenings in winter months, the Big Dipper seems to stand upright upon its own handle. This position is reversed in the summer months, when the "Great Bear hangs lowest in the sky" and "the Season of SED is that of the Great Night, when the Bear is slain, and this is in the Month of AIRU" (roughly corresponding to the zodiacal sign of Taurus—the sign of the Bull, as the Dipper is of the Bull's "Thigh"—and particularly to April 30, Cétshamhain or Walpurgisnacht). It would be possible to construct an occult calendar based strictly upon the perceived motions of this constellation and ignoring entirely the phases of the Moon and the seasons of the Sun, were it not for the fact that all of these motions are interrelated in the macrocosmic clockwork of the heavens. However, for those pursuing the course of initiation outlined in the *Necronomicon*, just such an occult calendar is appended, showing the appropriate dates and times for beginning the mystical Quest and for entering the Gate.

It is within the confines of the Northern Dipper that the Three Great Ones reside: three separate, but related, powers that are vulnerable to the invocations of the sorcerers. It is also the Northern Dipper that bears the strongest relationship to one of Aleister Crowley's most powerful god-forms, and one that is intimately connected with Sumeria and the cult of the *Necronomicon*.

Babalon.

Babalon is represented in the symbolism of Thelema as a Seven-Pointed Star, one point for each of the seven letters in *Babalon*. As we know, the stars do not have points, *per se*. They are objects similar to our own Sun, which as any schoolchild knows is itself a star, and nothing more. The author's contention is that such a concept as a *seven-pointed* star refers not to the number of sharp edges a star has, or to the number of rays it emits, but possibly to the number of *stars* in a particular star group, or constellation.

The Northern Dipper has seven stars in its constellation. So, of course, do many other constellations. What becomes important to understanding the link between Babalon and the Northern Dipper, however, is simple gematria, the system of numerology used by qabalists and magicians.

The number of Babalon (taking each letter as a number and adding them up) is 156. Crowley deliberately spelled the word this way, rather than the more common "Babylon." His reason for this lies in his attempt to align the concept of the Whore of Babylon of the Biblical Apocalypse—the consort, after all, of the Great Beast—to a more potent symbol. The number 156 is also the number of Zion, the Holy Mountain of the Qabalists, similar to the Mount Sumeru of Indian mythology, the Mount Kunlun of Chinese mythology, and the sacred "mountains"—the ziggurats—of Babylon itself. Zion, like Mount Sumeru, is perceived as the center of the earth, and of the entire cosmos by extension. (Parenthetically, the famous Tower of Babel at the city of Nebo in Babylon—now known as Birs-Nimrud—was exactly 156 feet high.) It is the pole around which the rest of the universe "rotates." Its identity with the Pole Star and the constellation with which the Pole Star is continually associated in the Asian literature cannot easily be discounted. The seven-starred Northern Dipper is the outward manifestation of the seven-pointed Star of Babalon. And these were represented in the Babylon of old by the seven-stepped mountains, or ziggurats, of the ancient Sumerian faith. (Indeed, one could make a case that the words "zion" and "zayin" are homophones and represent a play on words between Mount Zion and the Hebrew letter zayin, which serves as the number "seven"; in which case Mount Zion becomes Mount Seven. Just to confuse the issue a little, the Celtic word for Halloween—Samhain—is pronounced "Sah-ven.")

There are, however, antipodes to these seven stars, and they are the Seven Annunaki, "Lords of the Underworld, Ministers of the Queen of Hell." These might be considered as dwelling *below* Mount Sumeru, below Mount Zion, below Mount Kunlun, below the Mayan Temple of the Sun, in the Seven-Gated Underworld of which all the ancient texts warn us. They are the mirror images of the Seven Stars of the Northern Dipper, and represent the Nadir of the Macrocosm. They are referred to as "Dog-faced" in the *Necronomicon*, and this may possibly be a reference to the binary star Sirius, also called the Dog Star, and the subject of much controversy concerning the African Dogon people and their worship of that Star.[1] However, the dog as symbol of the Underworld is so pervasive in so many different cultures that one has to wonder which came first, the Dog or the Star's name?

The *Magan Text* is concerned with the story of the Descent of the Goddess Inanna (Ishtar) into the Underworld, a process that also consists of seven steps or stages, and which acts as an "anti-universe" to the Seven Gates of the *Necronomicon* itself. Paradoxically, one could not expect to ascend the seven celestial gates until one had descended the seven gates into the Underworld. This belief is reinforced by images from the Tibetan *Book of the Dead* wherein the first seven days after death the soul visits (or is visited by) demonic powers, and in the subsequent seven days visits angelic powers. The seven gates of the Underworld and of Heaven are thus dramatically illustrated: gates in space *and* time.

Maspero shows a figure on page 659 of his *Dawn of Civilization* that is of the goddess Ishtar "holding her Star before the Sun." This "star" is in reality a scepter, a long, straight rod with the inverted crescent at the bottom typical of *Egyptian* scepters, and surmounted at the top by a seven-rayed star. It is nothing less than the famous crown of Seshat that we have already described. To find it here, in the hands of the goddess who descended into the Underworld and representative of "her star" is evidence indeed that there is a celestial source for the "number seven" that we find so frequently mentioned in so many cultures whenever reference is made to the realm of the dead.

On an Egyptian papyrus from ca. 1400 B.C.—reproduced in Jung's *Man and His Symbols*—we can see the Egyptian Underworld pictured as a maze or labyrinth of Seven Gates, which Jungian author Marie-Louise von Franz relates to the labyrinths of European mythology, including that of Chartres Cathedral.[2] The entrance to the Malekulan Underworld is blocked by a Devouring Ghost tracing such a labyrinth in the sand.

Anu, Enlil, and Enki—or the Three-in-One of Daoist alchemy—are the primordial gods of creation, representing subtle forces and influences of which modern humanity has lost track, or calls by other names in a clinical distancing suggestive of a kind of psychic prophylaxis. It is possible that these three forces can be equated to the Id, Ego, and Superego of the Freudians; or the Shadow, Anima, and Self of the Jungians; or any other available and convenient trinity. The attempts of modern psychologists to frame such trinities shows how pervasive the belief in a primordial, preconscious group of "three" has become. Enki, Lord of Magick, in his underwater kingdom located just "above" the Underworld, may represent the subconscious as opposed to the unconscious, for instance; or he may simply represent the libido, where the Underworld itself is representative of the serpent brain, the "old brain" of modern human beings wherein the body's basic survival mechanisms may be found. This is material for a later study, but mention of it now may be useful to those readers who view occultism from a psychological angle. That the trinities of the various schools of psychology were prefigured in ancient texts—not to mention the most famous contemporary trinity of all, that of the Catholic Church—is worth the serious attention paid to it by the Jungians.

To the Daoists, the Three-in-One are the gods of the three "cinnabar fields" in the human body who exist simultaneously in each person and in the Northern Dipper itself. The cinnabar fields are the subtle centers most directly connected with the inner alchemy of Daoist practice, the areas of the body that assist in the purification and transformation of the entire organism. They are roughly analogous to three of the chakras of Hindu yoga.

The Chinese also believed in the existence of certain "antistars," stars that could not be seen with the eyes but that they believed existed as counterpoints to the visible stars. Hence, they posited the existence of an invisible Dipper around the visible one; they also went so far as to create complex formulas for determining the positions of antiplanets: an "anti-Jupiter," for instance, that is still in use today in Chinese astrology. They also posited the existence of two invisible stars in the Northern Dipper constellation, which are extremely important to Shangqing Daoist practices.

(Strangely, the Neoplatonists—particularly Proclus—also proposed the existence of invisible stars and invisible planets around the same time the Daoists were actually charting their presumed positions and invoking them with rituals.)

Astrology brings us to the next important point concerning the Dipper. Although the position of the Dipper at certain times is considered relevant to Chinese astrologers, none of the planets or luminaries ever "pass through" this constellation the way they do vis-à-vis the Zodiac. Instead, the Dipper rotates in the sky around the Pole Star, and acts as a kind of celestial "hour hand." Note was taken of the position of the Tail of the Bear (or Handle of the Dipper) at various seasons, and from this the actual hour of the night could be determined. Conversely, if the Tail pointed in a certain direction at a certain hour, one could determine the season. Silly as *that* sounds, it is relevant to our case of determining the precise moment the Gate is opened.

The Northern Dipper: A Biological Analogue

Just as the Pole Star was considered the northernmost point in the universe (at least, from the point of view of the Earth, which was considered the center of the universe at that time), and the top of a mystical mountain that had its base below the Earth itself, so was the same Pole Star perceived as symbolic of a center of energy that formed the topmost part of the astral body. In this way, it is similar (some would say, identical) to the thousand-petaled lotus *chakra*, the opening of which is the goal of kundalini yoga. To take this analogy a step further, the sushumna pillar (a path of energy in the astral body roughly analogous to the spinal column with its bundles of nerve endings, or to the path of the autonomic nervous system) of kundalini yoga can be seen as another manifestation of the mystic Mountain of the Chinese adepts. Indeed, Daoist "yoga" is itself concerned with opening a "mysterious gate" in the body corresponding to the Celestial Gate above: an *entrance* to immortality. This gate cannot be opened until the three cinnabar fields are brought into play. Cinnabar, of course, is one of the most important substances in Western alchemy as well as in the Chinese version. Composed of mercury and sulfur—two of the three basic elements in alchemy, along with "salt"—cinnabar represents a blending of the two mutually opposing polarities of existence that must take place before the Gate can be opened. (This is reflected in the statues of the God and Goddess that must be present during the rituals of the Walking in the *Necronomicon*.)

The seven stars of the Northern Dipper each have their own, spiritual, analogue in Shangqing Daoism. These may be likened to the seven chakras of kundalini yoga, and to the seven Gates of the *Necronomicon*. Indeed, just as the Gates of the *Necronomicon* each have a planetary analogue, so do the seven stars of the Northern Dipper. The five planets and two luminaries familiar to astrologers were believed to be the *grosser* forms of the seven principal stars of the Northern Dipper. And, as mentioned before, just as there were invisible "stars" forming an invisible "Dipper," so are there invisible "planets" forming an invisible "solar system."

We take this concept a step further by observing that, just as we have physical "visible" bodies, so do we have astral, or "invisible," bodies. The rituals we employ to summon the forces of the stars also work to summon the invisible forces of our own bodies, our own "souls."

It is important to practitioners of the *Necronomicon* to understand that what they are doing when they perform these rituals can have effects on several planes at once: the mental, the physical, the astral. In fact, one may not be too ambitious to say: the personal, the collective, the universal. By acting on oneself, one acts upon the rest of the World. By acting on the World, one acts upon oneself. The Eastern doctrine of karma is not simply a moral lesson; rather, it is the logical extension of the Hermetic axiom "as above, so below": the key axiom of all magick and, indeed, of all action whatsoever. By attempting to open the "mysterious gate" in one's own soul, one also risks opening the "Gate to the Other World" and thereby loosing demons upon the Earth. Such was the outcome of Inanna's return from the Underworld, as she reentered society trailing demons fore and aft.

This is the message of the *Magan Text* and, just as it was assumed that the practitioner of the *Necronomicon* rituals was a male, so was the descent into the Underworld symbolized by a female, a goddess, thereby implying a kind of cosmic polarity in which one could not exist without the other. And, just as the rituals of the *Necronomicon* may be employed by persons of either sex, so also we shouldn't overemphasize the gender of the person responsible for descending into the Underworld: after all, in the Christian era, Christ performed the same function in the death and resurrection mythos associated with that cult, via the identical symbol structure of being crucified (Inanna was tied to a stake and killed) and rising on the third day (Inanna was "dead" on the stake for three days). What *is* emphasized, however, is the very polarity of the universe as depicted in the *Necronomicon*: the two seven-staged processes are mirror images of each other, identical to the visible Dipper and invisible Dipper of the Daoist magicians. We will have recourse to this concept again and again in dealing with the initiatory structure of the *Necronomicon*.

The Great Bear Motif in Celtic Mythology

The Latin word "Ursa" means "Bear." Ursa Major is the Great Bear constellation that we have been discussing. Two of its stars form a line that points directly at the North Star, also known as Polaris or the Pole Star or the Lode Star. Polaris sits atop an imagined tent pole in the middle of the earth, and the rest of the stars in the sky depend from it as if on strings. In many ancient cultures, including the primitive, megalithic cultures of all continents, it was believed that climbing a tree or a pole during an initiation ceremony was tantamount to rising up towards heaven. This concept from before recorded history survives in many popular practices and cultic rituals today, including the May Pole of Europe and the *Poteau Mitan* of Haitian voudoun. The central pole is the *axis mundi*, a shaft that runs through the very center of the earth to a point deep in space. It is, of course, magnetic, which is a peculiar property of the poles. The very word "ursa," however, lends itself to a great many more revelations concerning the Bear and its occult significance.

Ursa comes from an Indo-European root, *rtko*, which means "bear," but from which a strange collection of other words also descends. Included is the Greek *arktos*, from which we get our English word *arctic*, not to mention *Arcturus*, the name of a star that is associated with the constellation *Argo* and of which R. K. Temple in *The Sirius Mystery* makes much without realizing that Arcturus once belonged to the Great Bear constellation in ancient times, as was known to the Chinese.[3] The association of the word for "bear" with concepts such as the Arctic (the northernmost part of our planet) and Arcturus, and from thence to the Argo of Jason and the Argonauts (and the Ark of the Covenant? And the Arcadia of the Masons, the *Et In Arcadia Ego* of the Cathar/Templar Holy Grail nexus?) would take us too far afield in this study, and readers are directed to *The Sirius Mystery* for a more detailed discussion of this "mythologem." For now, let us look at one last member of the *rtko* family, for in the Celtic language *rtko* becomes *arto*, and from there to the Welsh *arth*, from which the name "Arthur" is derived.[4]

Probably no myth is more central to Western European mysticism than the legend of the Holy Grail and King Arthur. It is not generally known that the name of Arthur means "bear." As the central figure of the Knights of the Round Table, his reign was seen as the pivotal moment of British history, when all the warring tribes of Britain were united under a single command for the first time. Arthur is said to have defeated twelve rebellious princes and to have fought twelve great battles against the Saxons. It was long thought that the idea of a King surrounded by twelve Knights was a Christian motif, like Christ and his twelve Disciples. However, if Arthur is a symbol of the Bear, might not the Knights of the Round Table refer to the twelve signs of the Zodiac, revolving as they do (in a circle, hence "round") about the central figure of the Bear, the most important circumpolar constellation?

(This points to another ancient Celtic belief, that of a *Dea Artio*: a goddess of the Bear. A sculptural arrangement at present in a museum in Berne, Switzerland, shows the goddess seated next to a small pedestal on which a basket of fruit is resting, facing a large bear who is backed up against a tree, as if bound to the tree.[5] This is another representation of the idea of a bear "attached" to a tree or pole, a northern European way of describing the Bull of Set tied to a stake, which is the celestial pole.)

Among the Chinese, the term "round table" has a definite

connotation. It's the heaven plate of the diviner's instrument, which rests on a square plate representing the earth. To a certain secret society today, the Round Table is a very complex system of correspondences, all of which are celestial in nature. Together with the Square Table and the Long Table, the three interlocking systems contain a Qabala that is superior to anything else in existence.

And, at the center of the Chinese "round table," what do we find but the Northern Dipper itself, alone of all constellations, either inscribed on the plate as a magickal glyph or present in its very essence as the magnetic needle. The Bear as center of the Round Table. Arthur and his Knights.

If we take the similarity even further, we are forced to realize that, according to the legend, Arthur's kingdom Camelot had become mysteriously barren. An object—the Grail—had to be found to revitalize the land. Knights went in search of this object far and wide. It was discovered in the castle of a mysteriously wounded (castrated?) person, the Fisher King. (Isis, after searching far and wide through the desolated Egyptian landscape for her husband—slain by Set—found the body of Osiris in the form of a tree built into the castle of a king of Lebanon. The Fisher King lived in Spain. Both were nearby foreign countries to the ravaged kingdoms in question. That is, they were "beyond the realm.")

The similarities we find in all of these myths are too numerous to investigate in detail. We will restrict ourselves to only those elements of the myths that directly concern our study of the Gates.

One thing is probably certain: King Arthur was as much a corporeal form of the Dipper as was Set. The historical romances built up around the Arthur legend—from the Welsh *Mabinogian* and the poems of Chrétien of Troyes and Mallory, as well as the *Parzival* of Wolfram von Eschenbach—include an interesting variation: at his death at the hands of his own nephew, Modred, Arthur is whisked away on a boat by angelic hosts to live forever in Avalon, awaiting the day when he will return to rule Britain. This idea of *some type of immortality* is a constant theme in dealing with the Dipper and its myths.

Another interesting resonance: Set is defeated by *his* nephew, Horus, and placed upon a boat—the eternal barque of the Sun God—where he stands forever, guarding the celestial vessel against attacks by the ancient Serpent, Apophis.

As the Cult of the Dead became more and more sophisticated in Egypt, the idea of the dead king rising to the stars after his journey to the Underworld became commonplace. And, of course, it would be a celestial boat that would take him there. In the Egyptian rite of the dead, this ascent into heaven could not take place until the mummy was touched by the sacred adze, the Thigh of Set, the iron instrument that was fashioned in the image of the Dipper. Need we wonder, therefore, to which stars these kings were believed to go?

The Celestial Calendar

The timing of the ascent to the Dipper, the Celestial Gate, is important. After death, in cultures as diverse as Tibet and Egypt, there is a certain waiting period (a period of *time*) that it is necessary to observe before the spirit of the dead individual can be expected to achieve immortality (or become reborn). Thus, the concept of the Underworld is both that of space (it is a specific location: an underground cavern or a mysterious fortress) and of time (the forty-nine—7 x 7—days of Tibetan Buddhism, for instance, or the seventy—7 x 10—days of the mummification process).

This space-time quantum is reinforced in the instructions we find in the *Necronomicon*, and in no other place is it explained so clearly. While pagan religions in Europe actually observed the same calendar as that employed in the *Necronomicon*, there is no explanation given as to why those dates are *ritually* so important.

"When the Great Bear is slain . . ."

"When it hangs from its tail in the sky . . ."

"The month of AIRU . . ."

At the time the *Necronomicon* was being written (not to be confused with the time those rites actually originated, in ancient Sumer), this period of time was—and still is—in the (tropical) sign of Taurus, the sign of the Bull; specifically, the above instructions refer to the pagan festival and Witches' Sabbat known as Beltane, or Walpurgisnacht; the Cétshamhain of the Celts: May 1st.

To the European Celts this marked the time of the lighting of great bonfires. The Celtic word "Beltane" is derived from two words: *tane* or *tene*, meaning "fire," and *bel*, a famous term—a survival of *Baal*—coming to the Celts a long way—from Syria and the Palestinian coast—and meaning "lord." Thus, the Celtic ceremony is one dedicated to the Lord of the Fire, or, perhaps, to the Shining Lord. The popular associations for an agricultural community were with fertility, for it was the time of Spring and of the planting. Rites were performed on May Eve to ensure good harvests and abundant crops. It was also the feast day of St. Walburga, an English missionary nun who accompanied the Benedictine monk St. Boniface to Germany on a mission of conversion, where Boniface became the first Bishop of Germany. (He is known as the Apostle of Germany.) Walburga outlived Boniface by about twenty-seven years, the latter having been murdered—together with thirty of his friends—in 750 by a mob of peasants (thus demonstrating a certain rough pagan élan when it came to dealing with meddlesome Christians?). These days, however, May 1st is no longer known as a feast day of St. Walburga, but has become replaced by one of the Feasts of St. Joseph, he of the flowering staff. In a related context, another pagan festival—Lammas—celebrated on August 1st has become the Feast of the Chains of St. Peter, commemorating the chains that bound the First Pope in his cells both in Jerusalem and in Rome (he brought his own?), which chains are still on display by order of Nero (!) in the basilica of St. Peter ad Vincula.

Much has been written elsewhere concerning the Beltane celebrations. Sir James Frazer, in *The Golden Bough*, has many pages devoted to its study. Several points strike us at once, however, about the form these rituals took, and that deserve some mention here.

In the first place, we should be careful to separate the purely *Pagan* celebrations of Beltane from the later Christian rituals that were designed to protect the countryside against witchcraft. In many instances the form is the same but the content has shifted to reveal a Christian, anti-Pagan bias. Quite often, effigies of "witches" were thrown into the great Beltane bonfires, but these were not originally of witches. Rather, in the pre-Christian, Druidic days human victims were chosen either by lot or by some other method and cast upon the flames as a sacrifice to ensure fertility of the fields.

In some cultures, two fires were lighted (always on a prominence, a high hill in the neighborhood, and never in a building or on top of a building), and those victims chosen by lot had to run between the fires three times. In other cases cited by Frazer, farmers drove their cattle between the two fires to ensure a good supply of milk, etc. The twin fires form a Gate Perilous through which one must pass to achieve—if not immortality—then a longer lease on one's mortal life.

In some parts of Europe, May Eve was greeted with a great deal of sexual license in the open fields during the night, and when daybreak came, the merrymakers would return to their village with branches of trees or with the May pole itself. In Tyrol, Beltane is a time when witchcraft is believed to be at its height, and in order to protect the village against the evil forces of decay and death (as they are perceived), the men of the village make a great noise with horns, banging pots, whistles, etc., and run around the village *seven* times.

This caution against witchcraft, of course, is merely the residue of a fear of Paganism that was instilled by the

Church. The Pagan Celts and other European peoples celebrated Beltane from ages past. The Church identified this practice as evil, and of the Devil (a Christian creation), and proclaimed all its adherents as "witches." Yet, there is no need in trying to deny that Beltane was a time of human sacrifice among the Druids and other Celtic peoples of the British Isles, a time when human victims were thrown—living—atop huge fires that were lit on the summits of hills to attract the attention of the gods. The spark that lit these great fires was believed to descend from heaven, from the abode of the gods, and at night when the stars are visible. The fires were lit with pieces of flint or cherd, for these produced a spark that would appear to be a bolt from the heavens. (This same practice obtained among the Aztecs, as well, and their sacrificial knives were made of the same substances.)

Sexuality, fertility, and human sacrifice. Witchcraft, drunkenness, dancing, and magic spells. Frenzied sabbats on mountaintops, and the May Pole that reached from the center of the earth to the highest point in the sky. The triumph of the Id over the Super Ego, if only for one day, one night.

In these agricultural communities, Beltane was the festival of the war between the Winter King and the May Queen, the time when the Old Man of Winter was finally slain—went to the Underworld—and the Queen rose from the dead. It was a time of reversed polarities, from the purely male to the purely female. To celebrate the planting of the crops, one could say that the "seed"—the male element—had been "buried," only to rise again at harvest time. And all this took place in the zodiacal sign of the Bull, and at a time when the Bull's Thigh had attained its proper position over the ritual site.

The Slaughtered Bull

The bull was a universal symbol of sexual potency, in cultures as diverse in custom, language, and era as those of Egypt and the British Isles. Indeed, our English words "bull" and "phallus" share the same Indo-European root, the word *bhel-*[2] which means "to blow, swell," and a host of other words having reference to "tumescent masculinity," as the *American Heritage Dictionary of Indo-European Roots* tells us. Another interesting derivative is the word "buttock," and possibly the word "baleen" for whale. (One is tempted to infer that the same root gave us the word *beltane*, but there is as yet no evidence for this etymology. The idea that *beltane* may be the time "of increasing [swelling] fire," however, is provocative. The possible rendering of "Fire of the Bull" is even more pertinent.)

The sacrifice of a bull was central to the cult of Mithra, which once vied for supremacy with Christianity in the Roman Empire. The famous tauroboleum in Rome attests to this: on the site where St. Peter's Basilica now stands, a platform had been erected for the annual sacrifice. A bull was slain over a grate, below which the new initiates would stand until they were covered in the bull's blood in a kind of baptism. So important was this concept to Mithraism, for instance, that Mithra is always shown plunging his knife into the neck of a bull. Mithraism may have even reached China, where it is said to have influenced the astrologers to such an extent that they used an abbreviated form of the word "mithra"—*mi*—to mean "Sun."[6]

In ancient Ireland, the method of determining who the next king would be involved the sacrifice of a bull. In this ritual, a designated individual would feed on the flesh of the bull and "drink his broth," after which he would fall into a sleep during which four Druids would recite incantations over the somnolent body. Whichever person of whom the man would then dream would become the next king. And at the base of the famous Gundestrop cauldron in Sweden there is a depiction of a hunter killing a bull, attended by two dogs. Historians have been at a loss as to how to attribute what must be a very important ikon, due to the lack of written or other records incorporating this motif. With modern archaeology pointing to a possible Indus Valley origin for this cauldron, however, we are on firmer ground in insisting that it reflects the Mithraic Cult of the Bull very clearly, even to the inclusion of the dogs that are Mithra's constant companions.

That fertility rituals often include scenes of unspeakable horror and cruelty is common knowledge to anthropologists everywhere. The fertility ceremonies of certain parts of India, notably Bengal, involved fearsome rites of human sacrifice in which the blood of the sacrificed victim was sprinkled on the fields to ensure fertile land and abundant harvests. The Aztecs, as is well known, offered up thousands of victims to their gods as a guarantee of rain for their fields. That fertility and death are somehow inextricably connected is emphasized in the Celtic ritual of the Maypole.

The Celestial Pole

On May 1st, after the witches have gathered on the summit of Mount Brocken for their frightful sabbats the evening before, and after the bonfires were lit all over the countryside for leaping over and for passing children through to bless them and cure various ills, the people would meet in the daylight and dance around a gaily decorated pole set upright in the ground, made from a freshly felled tree. People danced around the pole in what appeared to be another fertility rite with the pole signifying a phallus.

However, this explanation may fall short of the real meaning of this custom.

While fertility poles are common throughout the world from India to Ireland, the modern belief that they are merely replicas of divine phalluses, pure and simple, is an example of a kind of pseudo-Freudian shortsightedness. For, everywhere we find a "fertility pole" we find some reference to the heavens: to a source of divine energy or potency that is believed to permeate the entire universe from some cosmic source. The equation of peril and fertility with the rites of the Beltane bonfires speaks to us directly of the shamanistic

Quest. The equation of a tall pole erected in the center of the town about which a round-dance takes place speaks to us equally directly of a primitive simulacrum of the dance of the stars around the celestial pole. This same Pole figures prominently in the rites of the Malekulans, the Siberian shamans, and even of the modern practice of Haitian *voudoun* in the central emblem of the *peristyle*, the *poteau mitan* before which ecstatic dances and spiritual "possessions" take place to this day. In Haiti, the gods slide *up* the *poteau mitan* into the sacred arena to mingle with the celebrants. In either case, the central pole of rituals as diverse as the May Day revels of Europe and the midnight ceremonies of Haitian *voudoun* is a means of connecting the world of humans with the Otherworld, the Underworld of the primal gods whether it is perceived to be below the earth or in the heavens. The pole points to a *space*—either directly below the ground, at the nadir, or directly above the ground, at the zenith, the twin poles of the earth's axis and the celestial axis combined into one—that can be entered only at a specific *time*.

At Samhain—or Halloween, as it is known in the States—the spirits of the dead visit *us*. They open the door from their side. The sign of the Bear, of the Bull's Thigh, stands upright in the evening sky. But on Cétshamhain—the "anti-Samhain" feast of Beltane ("Bull's Fire"?) in the sign of Taurus, the Bull—we can open the Gate in the other direction and take the first steps down into the Otherworld.

And at just before midnight in the evening of April 30, in the sign of the Bull, the Great Bear (the Bull's Thigh constellation) hangs from its tail in the sky . . .

CHAPTER 4: The Great Bear in Shangqing Daoism

Much of what has been written in ancient times about the Great Bear—and when we speak of the Great Bear, we mean specifically that asterism composed of only seven stars called the Big Dipper or the Northern Ladle—is preserved in the classical Chinese works on alchemy of the Mao Shan School, from the Jin Dynasty era of about 265-420 A.D. We have great evidence for the importance of the Great Bear in ancient Egyptian mythology and ritual, where it is identified with Set and with iron, and with the orientation of the temple. The Thigh of Set is represented in the ritual by an adze—or a pair of adzes, called the "celestial iron scissors"—a piece of shaped iron that was used in the mummification rites along with the Ur-Hekau to "charge" the corpse so it would live again. It was used to open the mouth and eyes of the body, to awaken the senses. Sometimes this process was accomplished by Ptah, the Lord of Creation—that is, by his human representative—who held the adze or adzes.

Set, as a constellation, was pictured as a bull, tethered with two ropes to a stake and revolving endlessly around Isis, who was pictured (in her manifestation as Ta-Urt) as a hippopotamus standing upright next to the stake. Isis in another manifestation is also the Dog Star, Sothis or Sirius, on whom the Nile and the entire life of ancient Egypt depended. Osiris corresponded to our present day constellation of Orion, the Hunter. We have here the three symbols that are found on the Gundestrup Cauldron: the Hunter, the Bull, and the Dogs. If we remember our Egyptian religion, we know that Osiris was killed by Set, buried alive in a coffin. Isis was able to find Osiris's body—at Byblos, in Lebanon, where the coffin had become part of a tree that was later cut down and made into the pillar of a royal house (in a later cult, Christ referred to himself as the cornerstone of the temple)—and magically impregnate herself with his semen and thereby give birth to the avenging son and magickal childe: Horus.

Set, enraged at this maneuver, then found Osiris' body and tore it into fourteen pieces, which he scattered over all of the known world. (The number "fourteen" may refer to the seven Gates down into the Underworld plus the seven Gates up into the Heavens.) Isis traveled throughout the landscape, finding piece after piece and building shrines where they were found, until she had located the entire body . . . except for the phallus. The phallus had fallen into the Nile and was swallowed by a fish. This phallus has often been referred to in occult literature as the "Talisman of Set."

(There is an interesting parallel here to the Roman Catholic ritual of the Fourteen *Stations* of the Cross: fourteen specific sites around the interior of the church where the arrest, trial, and crucifixion of Jesus Christ are remembered. Christ, like Osiris, is a slain and resurrected God, betrayed by a friend, whose torn body is a focus of worship. The Fourteenth—and final—Station shows Christ being placed in the tomb.)

There seems to be some connection between this "talisman" and the mysterious Thigh of Set, for they each serve a resurrecting function, are both phallic, and are both assigned a correspondence to Set. The talisman of Set is so-called because as long as the phallus is not assembled with the rest of Osiris' body, Isis can give birth to no more avenging deities. The phallus of Osiris is, then, a talisman preserved by Set in a secret place to guarantee his own, personal, immortality. Indeed, the Great Bear constellation itself is referred to as "the Indestructible" for—in the northern latitudes—it never disappears below the horizon but revolves endlessly around the celestial north pole. This principle of immortality as related to Set—and *not* Osiris—is something of a paradox until it is realized that Osiris was slain and resurrected, but Set *never dies.* We are obviously speaking of two different concepts of immortality when we speak of Set and Osiris, who were brothers. It is just this twin aspect of immortality that most concerned the ancient Chinese sages, and they, too, assigned an important role to the Great Bear constellation and the shaped piece of iron that they used in their own rituals and for orienting their own temples. Perhaps we can obtain a greater insight into the importance of the Great Bear and its relevance to the rituals of the *Necronomicon* if we take some time to investigate the amazing parallels between the form of Daoism known as *Shangqing* with its "talisman that opens the Celestial Gate" and the mystical instructions of the Mad Arab.

The term *shangqing* is translated as the "Heaven of Great Clarity" or the "Clarity Above," meaning above the Nine Heavens of ancient Chinese cosmology. In this sense, perhaps, it can be perceived as the sphere of Kether in the Qabalistic Tree of Life. It is called the Abode of the Great One, the Father of the Dao, who is older than heaven and earth. The term "Nine Heavens" refers to the seven stars of the Northern Ladle plus two "invisible" stars, *Fu* and *Bi*.

As a system of Daoism, Shangqing can be identified as far back as the fourth century, A.D. This was roughly the same time Christianity was becoming politically consolidated in the West. The *I Jing* (I Ching) was already ancient, Buddhism had existed for nearly a thousand years, and astronomically oriented temples were being built in North America by persons unknown. The religion of Islam had yet to be created by the Prophet Muhammed. And the Jews were in Diaspora, Herod's Temple having been destroyed over two hundred years previously.

Astronomy as a science was already well known among many civilized peoples of the world, including the Chinese, who had a system of astrology quite different from the one with which we are all familiar in the West. Central to their astronomical/astrological computations and observations was the Northern Ladle. It was used to "set the four directions" and to divide the year into seasons, and the night into hours. It was the Chariot of the Celestial Emperor, who was believed to ride around the Pole Star continually, overseeing his earthly kingdom. (And today, the motions of the stars are used to set our sidereal clocks, which in turn are used to set the clocks we all use every day.)

Originally, the Northern Ladle was composed of *eight*, not seven, stars. The missing star is Arcturus, which once formed part of the Dipper over three thousand years ago. Now, Arcturus is no longer a member of the circumpolar society and drops below the horizon for long periods each year. Of course, veteran star-gazers will know that the Big Dipper *does* contain an extra star—Alcor—that is not always visible to the naked eye and is located exactly where Daoist tradition states one of the two "invisible" stars is located, as a companion of Mizar (the other would be a companion of Phecda). This brings us to another consideration, that of the occasional reference to "eight" stars or the "eight-pointed star" that symbolized divinity to the ancient Sumerians and forms the basis of several magickal glyphs in the *Necronomicon*.

To a Chinese sage or alchemist of the second century, A.D., a reference to the "eight stars" would have meant the Northern Ladle plus the Pole Star. After all, the Northern Ladle both rotates around, and points to, the Pole Star. The Pole Star can only be approached through the Seven Gates of the Northern Ladle, and represents the highest attainment possible for earthly beings. It was the central point of the diviner's plate, and the Chinese used the diviner's plate to divide the earth and the heavens in dozens of different ways. After determining the four cardinal directions, they then developed the "cross-quarter" directions, or northeast, southeast, southwest and northwest, for a total of eight major directions. Each of these directions was represented by one of the eight trigrams—or *ba gua*—familiar to most of us from the *I Jing*.

These eight trigrams were then arranged in a square around a central square that represented the mystical "center." The resultant figure is well known to Western occultists as the Square of Saturn, and to Qabalists as the table of *Aiq Bkr*, a numerological device for computing the values of Hebrew words. A glance at any of the modern texts on *I Jing* research that have been published in English in the last thirty years or so will demonstrate that the Chinese attached great significance to this "magick square" and it was involved in seemingly endless variations—some of which make no mathematical sense but are considered mystically very powerful—of the nine squares. In fact, the Imperial Palace itself was divided into a square of nine chambers, where sacrifices were performed in accordance with the seasons in the appropriate chamber.

By the time the peculiar form of Daoism known as *Shangqing* was established, the Cult of the Northern Ladle had achieved a very high state of development. The Northern Ladle was the central astronomical, macrocosmic *fact* around which the erection of temples, the arrangement of the calendar, the timing of important events, and the pursuit of one's personal spiritual evolution took place. In China, the Emperor, the Land, and the People were considered part of a single organism that had to work in harmony with the stars if it was to flourish. Although a great deal has been written about *feng shui* "geomancy," an important element of that art is often either overlooked or given minimal attention: that the "telluric" or earthly currents it is supposed to discover and to measure are a complex function of the earth's magnetic field, the position of the Northern Ladle, and the combined effects of the Sun and Moon. The many divisions and subdivisions of the diviner's plate (including the basic division into an "earth" plate and a "heaven" plate) bear this out. The combination of the ten Heavenly Stems and the twelve Earthly Branches are also indications that, while the forces of heaven and earth are different, and measured in different ways with a different numbering system, the two disparate forces taken *together* are what constitute the power and vision of the sage, geomancer, astrologer, or magician. Either one taken by itself is deficient in meaning. This central concept is often lost in modern discussions of astrology, where the accent is always upon the heavens. Naturally, our systems of houses, rising signs, declination, the ecliptic, etc., all point to an *admission* of the interrelationship that exists between the earth and the heavens, where sometimes the *understanding* of this relationship is lacking. An example might be the determining of the best *time* to build a house while not taking into account the best location and orientation (*space*) of the house-to-be. In Asia, *both* of these factors are involved in any important decision.

But to the Shangqing alchemists, this knowledge—while extensive—was not enough. It was not enough to simply regulate one's earthly affairs in accord with the position of the Northern Ladle. One had to unite oneself with that celestial power to attain immortality and discover the source of human origins and the direction of human destiny. One had to go there.

Modern critics of astrology and the "occult" sciences are always quick to ridicule the "superstitions" of the previous age when people believed there existed a complex relationship between events on the earth and in the heavens that could be expressed by numbers and diagrams. Numerology is denigrated; astrology scoffed at. As imperfect and vague as many applications of these concepts may be—and this author does not deny for a moment that they are—they nonetheless reveal an extremely scientific predilection among their adherents: the desire to express everything, all experience, in terms of number, the better to control the forces of nature. This is not superstition; it is the same urge that fuels all modern science, whether biology, physics, chemistry, or even psychology. The impulse to find a numerical means for expressing events as disparate as the motions of subatomic particles or the incidence of suicide among certain social groups is an indication as to how prevalent this view is in modern science. We can say that modern science *is* number; it is the glue that holds it all together. And so it was for the occultist, the alchemist, the astrologer.

For the modern world, number is the key to curing disease, prolonging life, and reaching the stars. It was just so for the medieval magician and the ancient alchemist. In those days, magick *was* science; alchemy *was* both medicine and chemistry. The difference is: they were both informed by an idealism, a spiritual zeal to use the elements of matter to improve the world and to perfect humanity. That impulse is sorely lacking today; we have to go back in time hundreds, even thousands, of years to learn what that type of emotion feels like.

And here is where we enter the mysterious realm of the Chinese alchemist/philosopher/magician; we will open the double-locked doors of the hidden chamber; we will shine a feeble light on the silk cloths embroidered with the signs of the constellations; we will gaze in wonder at the vision of the dancing stars and the dance *of* the stars, the strange "Pace of Yü"; we will stand in the center of the first ever recorded Rites of the Pentagram, and we will breathe the incense.

The Rite of the Pentagram, and of the Watcher

A ritual peculiar to the *Necronomicon* is the Rite of the Watcher. The Watcher "cares not what it watches," but is a protective force that guards the magician during ceremonies. Sacrifice must be made to it, and to neglect to do so means that the Watcher will turn on the magician instead. It is a blind force, and must be controlled; but potent, indeed.

The Watcher is summoned by a sword that is stuck in the ground. That is its symbol: the upright Sword. Yet, as far away in space and time as the Shangqing schools of China, we find the same ritual being enacted for the same purpose. Here, it is called *chien-chieh* or "sword liberated." A sword is used as a temporary "stand-in" for the body when the magician's astral body is transformed and vanishes into the Other World. The sword stands guard until the "body" returns.

The term *chien-chieh* is a pun on another term, *shih-chieh*, which means "corpse-liberated." Both are Daoist terms for "transformation." All of this refers to a lost Daoist work known as the *Sword Scripture*, which concerned "liberation" and transformation by means of a sword. What this means is not well known, except for the fragments that have survived in such concepts as *chien-chieh*. Also involved in discussions of the *Sword Scripture* is the Jade Elixir, a deadly poison that was supposed to transport the magician to the realm of the Winding Constellation (the Northern Ladle, or Big Dipper). There is no doubt that the consumption of this "elixir" was an act of ritual suicide; yet, immortality and transformation was its goal. The act of imbibing this toxic substance took place in a cave, at an altar facing north. The cave, the northern direction, the Winding Constellation . . . all of these are common motifs in the universal concept of descending into the Underworld; the Chinese simply took it one extra step, and incorporated an act of suicide to propel the magician directly into the Underworld while in a state of ritual preparedness. The elixir contained substances of "extreme astral purity"[1] that would prove poisonous to a mortal being, since the human body is a mere sheath of the astral body, composed of various perishable elements that cannot withstand the undiluted essence of the stars.

Naturally, this was not the preferred method of *all* Daoists, and we will gladly concentrate on alternate means of transformation and liberation. This method is introduced solely to acquaint the reader with the ritual of the Watcher as it was practiced by the Chinese masters in an almost identical form. It should be emphasized, however, for any reader foolish enough to want to go that route, that the adept had undergone an intensive period of training in Daoism that lasted many, many years under a capable master. The precise formula of the Jade Elixir has been lost, so it would be futile to attempt this ritual by any other means.

It should be understood by all that the urge to self-annihilation that is common to some beginners in occultism is a manifestation of the unconscious urge towards destruction of the *ego* and not of the entire organism itself. What is perceived to be the urge to annihilate the *self* is the first stage of the shamanistic Quest in which the self is confused with the ego, which is a tool of the self. When this confusion takes place, there is certainly the propensity towards various types of neuroses, and these were managed in shamanistic societies by those who had already "broken down" and returned, cured, to help those others who would find themselves in similar circumstances. It is not the intention of this author to criticize what seems to be an Asian taste for ritual suicide, but merely to inform the reader that even in those societies, other means towards the same ends were discovered and employed to great success, as they are routinely in the West. *Caveat lector*.

In Sumeria, the star symbol was always a sign of divinity. On the cuneiform tablets and cylinder seals, the placement of a star next to an image of a person meant that person was a god or goddess. Star = God is the earliest recorded theological equation.

In Egypt, the star symbol—when enclosed by a circle, like our modern pentagram symbols—indicates the Underworld. A free-standing star is, simply, a star in the heavens. The "inverted" pentagram—as a sign of the Underworld or of Satanic entities—was unknown to the ancient Egyptians. Yet, it is worthwhile noting that their symbol for the Underworld was still a star—a celestial body—even though it is mysteriously enclosed by a circle.

The source of the five-pointed star was Sumeria, where it seems to have begun its life as a symbol for the plough. It was called AR, or ARRA, and early commentators on Sumeriology seem to feel that this was the signature of the linguistic group that later came to be called "Ar-yan," i.e., "People of the Plough." The plough implies knowledge of agriculture, which is one of the arts communicated to the Sumerians by their mysterious benefactor, Oannes. That the plough symbol metamorphosed into a star symbol may be a kind of visual punning in which the initiated understood that their civilization owed its existence to the stars. In Egypt, the Pharaoh was said to become one of the stars after death, and from there to "plough the earth."[2] As a coincidence, the Dipper constellation itself is referred to as the Plough in the British Isles.

That the pentagram is supposed to possess great power over demons and other evil forces goes back at least as far as Vedic India, and is known in the West as part of the history of Solomon and the building of the Temple. According to the legend, Solomon built the Temple with the aid of demons, whom he later shoved into a bottle and sealed with his Seal, the Five-Pointed Star, and tossed into the sea. There they remained, until the bottle was washed up on the shore one day and opened by an unsuspecting Arab. To these demons the Pentagram is the symbol of their imprisonment and of the great Magician who put them there, Solomon. Hence, their fear of this sign.

It is interesting to note that the Pentagram we refer to as Solomon's Seal is indeed a five-pointed star within a circle, the ancient Egyptian hieroglyphic for Underworld. Also, the number of demons forced into the bottle is seventy-two, the number of the demons we find in the *Lesser Key of Solomon,* also known as *Goetia.* Multiply the number of demons by the number of points in the pentagram and you obtain 360, the number of degrees in a circle. This may be another form of ritual punning, and it is mentioned here for those who have the wit to make something of it. A plough, of course, digs up the earth and disturbs the Underworld, so there may be a connection there with the rites of Osiris.

The star rituals of the Daoists were more complex than their Western counterparts. They involved a combination of eight "effulgences" multiplied by the three cinnabar fields to obtain twenty-four "perfected immortals." These were arrayed on the vertical plane surrounding and enveloping the adept. Then there were the five "pearls" that represented the four pillars—or watchtowers—of the four directions and the center—or middle—pillar. These are arrayed on the horizontal plane, and symbolize the philosopher's stone, the "pearl of great price," as it was referred to in the West.

Three cinnabar fields, eight effulgences, twenty-four perfected immortals, and five pearls . . . altogether these elements array the adept like a robe of light, forming a separate body that will travel to the stars.

The Daoist ceremony we are describing is a perfect blend of yoga and magick, something that Aleister Crowley was at great pains to provide for his students and followers. He was considered ahead of his time in the West, but in the East these practices were known for nearly two thousand years before Crowley was born. The Daoist adept had to purify his body and train his mind in deep concentration before he could successfully manage the type of spiritual transformation we are discussing. An essential part of his meditations was the Northern Dipper, shown in the ikon of the Great One, the first God, depicted in China as a baby (Horus?) holding in one hand the Dipper by its handle (like an adze?) and in his other the Pole Star.

The adept would array his body in the stars of the Dipper, and there were even specific recommendations as to how this would be done: which star to appear over which part of the body, and when this meditation should be performed.

The room in which the ritual/meditation takes place must be double-locked. The ritual takes place at midnight, the hour when the yang energy begins to rise. The adept has prepared by bathing and fasting. Incense is lighted, and the adept grinds his or her teeth a specific number of times (which varies from school to school). This summons the yang energy in the body. The adept invokes the four heraldic animals of the four cardinal directions, just as is done in the Golden Dawn rituals. Then, facing east, the adept visualizes the stars of the Dipper slowly descending, the handle pointing due east. This is the preliminary rite of protection, which serves to preserve the body of the adept during the remainder of whatever ceremony is planned.

A typical meditation would continue by turning one's gaze inward and by "seeing" oneself as if in a mirror. The eight "effulgences" are invoked and visualized in each of the three planes called the cinnabar fields. These correspond to the twenty-four perfected immortals, which are after all metaphors for internal conditions in the body mind continuum. They represent the twenty-four Solar Terms, the constellations of the year, and are basic forms of physical energy present in the body. This *identity* of inner, human thought-forms and biological "events" with the stars in the sky is constantly, consistently reinforced. It is the axiom on which all else depends. And, according to the Chinese masters, these exercises are best performed at the solstices and the equinoxes.[3]

At this point the small, dark, hermetically locked meditation chamber has become filled with the light of these shining orbs. The five "pearls" are made manifest, the light from the shining orbs illuminating the four quarters and the central (Middle, Sushumna) pillar. At that moment the radiance from this accumulated light lifts the adept upward, out of the frail, perishable body and heavenward to its natural abode among the stars, the stars of the Northern Dipper.[4]

One thinks of the hermetically sealed space capsules of modern astronauts who are also engaged in a project of travel to the stars. The darkness of space and the darkness of the meditation chamber. The silence. The realization that one is in danger, and that one is absolutely and utterly alone, with only the beings on the stars for company.

The great adepts who created the Golden Dawn could not have known about these practices, for they did not number a Chinese speaker or Daoist adept among their company. These texts are only gradually being translated into English and other Western languages, although they have been available for some time in the *Taoist Canon* published variously in Shanghai and in Taiwan at different times. Mathers spent all of his time in the British Museum, translating the books of the Qabalah and the works of Western ceremonial magicians, for which we are all genuinely grateful. Crowley, although he spent a little time along the Yunnan/Annam border, did not speak or write Chinese and was certainly not acquainted with any of the Taoist Canon other than what was already available in English at that time, namely the *Dao De Jing* (*Tao Teh Ching*), the *I Jing* (*Yi King* or *I Ching*), and other standard works. Daoist alchemy in general has been *terra incognita* until comparatively recently, with the publication of such works as those of Charles Luk and the spate of new books on inner alchemy, *qi gong*, and the *I Jing* coming out of Taiwan.

This goes some way to explaining Crowley's imperfect understanding of the trigrams and hexagrams that make up the system of the *I Jing*, and his awkward attempt to assign them places on the Qabalistic Tree of Life, or to otherwise connect them with planets and signs of the Zodiac. Although the attempt must be applauded, the results are not worthy of retention. As I say, this is not entirely Crowley's fault but is due more to the paucity of translated material on Chinese mysticism that was available to the West at that time. Had Crowley had access to the recently published exegisis on the works of the Mao Shan school of Daoism, he would have undoubtedly recognized them for what they were and incorporated them into the revised magickal system that forms the backbone of his secret society, the A . . . A . . .

The Ceremony of the Walking, and of the Circumambulation

This experience of rising on the ladder of light, of strolling among the seven stars of the Northern Dipper, is called the *Pace of Yü*. Yü was the Chinese Oannes, a being who mysteriously appeared and gave the Chinese their calendar, their occult knowledge, their basic civilization skills. He is a being who performed the first measurements and who gave the Chinese the "ur-chart": a prototype of the Ho Map, Lo Map, charts of the most ancient arrangements of the eight trigrams in magickal squares.

The "Pace of Yü" refers to another one of his gifts to humanity: the method by which one walks among the stars. In the *Necronomicon*, this is known as the Walking, and is concerned with entering the Seven Gates. In China this is also a Walking ritual designed to pace the seven stars, or gates, of the Northern Dipper. In this way the two rites from completely different cultures help to clarify and explain each other.

In the literature of the Golden Dawn and Crowley, this "walking" is symbolized by the circumambulations that take place in the liturgies of the various temple rituals. The Gnostic Mass is one example, in which the priest circumambulates the temple in a precise fashion. The meaning behind these sometimes intricate pacings has been all but lost with the passage of time . . . or possibly handed down orally among initiates. We can identify the circumambulations that are required in the Gnostic Mass by comparing the path of the circumambulation to that of the orbit of Sirius as illustrated in the Temple book already mentioned.[5]

The circumambulations that we can observe in nature are exclusively those of the heavens. In fact, if we are to interpret the term "circumambulation" tightly, we may insist that any such path be referred to those of the circumpolar stars about the Pole Star. The Pace of Yü is precisely that: an enactment on earth of the journey to the Northern Dipper, from star to star, beginning with the "first star"—the outer lip of the bowl of the Dipper—and continuing on to the last star and ultimate Celestial Gate, the star at the very end of the handle.

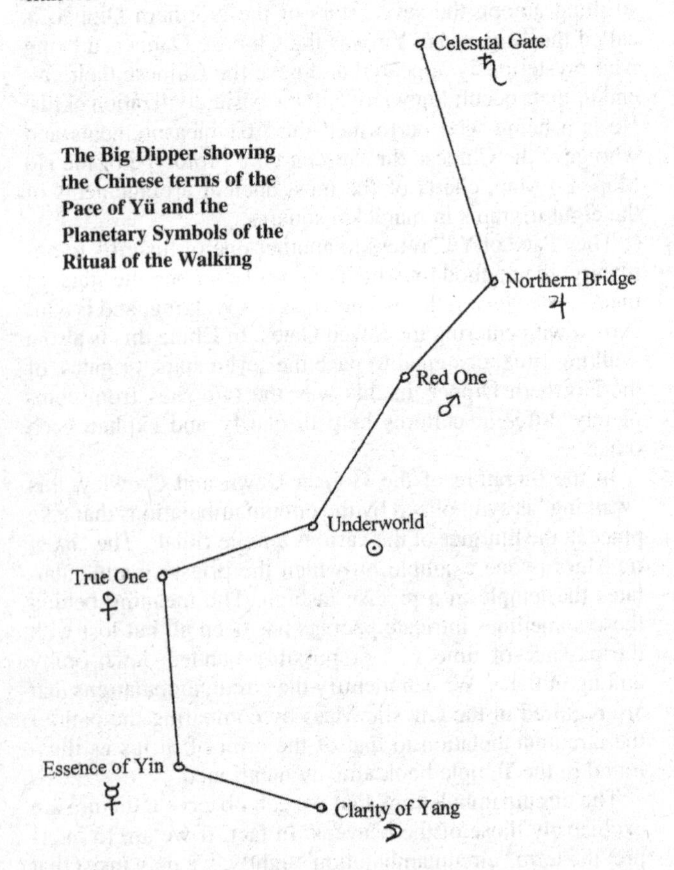

The Big Dipper showing the Chinese terms of the Pace of Yü and the Planetary Symbols of the Ritual of the Walking

The first star is called "Clarity of Yang," while the second, right below it on the bowl, is "Essence of Yin." As mentioned, the ritual of Walking the Dipper begins at the midnight hour, when the Yang energy is rising, and begins with the first star, the "Clarity of Yang." The fact that we begin with this particular star indicates once again that, for that star to be "nearest" to us, the Dipper must be hanging "upside down," i.e., from the tail of the Great Bear. The circumambulation of the Dipper is not a mere passing from one star to the next in succession; rather, it is a definite ritual with carefully defined steps, like a dance. The dance can be visualized during meditation, strongly imagining the body performing each of the prescribed motions in their proper order, or it can be danced within the sacred space of the meditation chamber.

Another method of "rising on the planes," of Walking, or performing the "Pace of Yü," consisted in lying down on a carpet embroidered with the stars of the Northern Dipper constellation and invoking each of the stars by name and with the appropriate words of power.

As we can see, the methods varied but the goal was always the same: the creation of an "astral" body, a body composed of the same pure substances as the stars, and rising from the earth to the heavens, to that specific constellation known as the Northern Dipper. This sense of direction was so keen that diviner's plates—complete with magnetic needle—were often buried with dead emperors so that the deceased could find his way to the constellation where he could take his rightful place among the stars. This was an idea not unique to China, but one that can be found in Egypt and in Mexico, as well.

Nor was this practice limited to solitary adepts. Groups of initiates could perform these rituals, and in some cases sexual intercourse was also incorporated into the general framework. These practices involved the cultivation of *jing* (*ching*) or the divine, transforming power that is carefully tended within the body of the initiate(s). *Jing*, although often visualized as a luminous essence or spirit in the initiate's body, was the same stuff as the stars, an astral fluid or power that was not *similar* to—or *approximated*—the astral matter, but was the *same as* the essential stuff of the stars. "The *ching* are essentially bodies of light."[6]

The "dual cultivation" (i.e., sexual) method was referred to as *ho-ch'i*, or a "mingling of sexual breaths."[7] This had to be performed under the supervision of a Master, and was the somewhat grosser form of *ou-ching* in which the adept—operating alone—achieved union with the "Celestial Maiden." In the former case, it was not uncommon to find that "dual cultivation" led to, or was accompanied by, sexual orgies of the most outrageous type, which resulted in many local scandals but whose nature has never been proven to be other than purely magickal and spiritual. One cannot always take the reports of horrified noninitiates in these matters at face value; the existence of orgiastic rites among the followers of Dionysus and in the fabled Rites of Eleusis does not detract from their spiritual nature, and indeed, such orgies were central to their entire purpose. The socially condoned sexual license of the Beltane celebrations is another case in which orgies have served noble—or, at least, practical—functions.

There was, of course, room for the ascetic who wished to follow the Way of the Master alone in the mountains. This was referred to as *ou-ying* and means the same as *ou-ching*. The essence is still cultivated, still "paired," but as in the *ou-ching* method, the partner is a "celestial maiden": another "body of light." The techniques of both *ho-ch'i* and *ou-ying* or *ou-jing* are sexual in that the energy—the *jing*—is the subtle (astral) basis for sexual energy. Both types of adept are quite concerned about cultivating the sexual energy in their own bodies and redirecting it towards magickal, spiritual ends. The difference lies solely in the physical practice, and the decision as to which path is the wisest must rest with each individual, or with that individual's Teacher.

With the spread of Confucianism, Daoism seemed to become less a "practical" philosophy and more a "moral" philosophy. Some authors suggest that this was the rational, benign influence of Confucius, which stressed familial loyalty and service to the Emperor. This author's belief is that the Daoist sages saw the writing on the wall. By the sixth century A.D. the Chinese Emperor would formally repudiate the religion and practice of Daoism in favor of Buddhism. Between the second and fifth centuries A.D., therefore, Daoism had grown from a small, disorganized cult into something approaching imperial status, only to find it all whisked away again by imperial decree. In order to survive, the Daoists had to take on the persona of the competition. This is so in keeping with Daoist philosophy concerning politics that this author cannot believe that the Daoists simply abandoned the practices of their masters, exchanging them for a sterile moral code replete with prayers for the Emperor's health.

Whatever happened, and for whatever reason, the practices of the Shangqing Daoists continued under cover in isolated temples and mountain retreats for a long time. In the twentieth century, Blofeld recounts several instances of discovering Daoist sages at work cultivating the inner essences and, in some cases, possessed of supernatural insights and power.[8] The survival of this Art in the East is paralleled only by the survival of ceremonial magick in the West, a survival that is due largely to the efforts of men like S. L. MacGregor Mathers and Aleister Crowley.

CHAPTER 5: The Formation of the Astral Body

What is an "astral" body?

During the last one hundred years the people who have popularized the notion of an "astral" body in the West have been the Theosophists and those organizations that claim a Theosophist lineage or tradition. Thus, we find the astral body mentioned in works by Rudolf Steiner of the Anthroposophical Society, in teachings of the Liberal Catholic Church (which shares a relationship to the Theosophical Society very similar to that enjoyed by the Gnostic Catholic Church to the OTO, a German secret society of the early twentieth century), and among the various writings of the Golden Dawn leaders and members. It is probable that Madame Blavatsky—foundress of the Theosophical Society—obtained this idea of an "astral" body during her lengthy sojourn in India. Indeed, much of her writing is concerned with Indian concepts (such as the chakras, the subtle body, various levels of awareness, a kind of spiritual "caste" system, etc.), so that it would come as no surprise to anyone that the astral body is only one of many such ideas to have originally come from her substantial intercourse with the East.

In Indian—which is to say, primarily Hindu—terms, the gross physical body is but one of several sheaths encompassing the human spirit. There is also an astral body. Authorities differ on the number of these sheaths and their nomenclature. To simplify matters, the author proposes that we study the Indian and the Chinese concepts in tandem since they both address the deliberate manifestation of one particular body, a kind of spiritual sheath or envelope, through various physical and mental exercises.

The very term "astral" refers, of course, to the stars, a point that is regularly overlooked in general discussions. The astral body is therefore composed of some substance in common with, and peculiar to, the "stars." It is the part of ourselves that derives from our stellar origins. If the myths of Egypt, Sumeria, India, and Mexico share something in common, it is that humanity was created from the body (flesh) of the Serpent and the spirit (breath) of the God or Gods who defeated the Serpent, and that these Gods came from the Stars. Our astral body, therefore, is that essential element of ourselves that is a direct legacy from the star god who created us.

Blavatsky and her followers made much of something they called the "astral plane": a level of existence beyond the purely conscious, three-dimensional world in which most of us are fortunate enough to find ourselves. The astral plane—the realm of the stars, in effect—is invisible to the ordinary eye, but its influence on events in our world is enormous. As we know, occult theory is full of such references, the very term "occult" itself meaning "secret, hidden." The Golden Dawn talked about "secret chiefs" who were the invisible—or only occasionally visible—leaders of the Order, a concept that is echoed in the idea of the Twenty-Four Perfected Immortals of Daoism.[1] The Islamic sect of Shi'ism speaks of the "hidden Imam," etc. We should address immediately what becomes obvious to the reader with a little thought: that the existence of a secret plane of existence where all the events of our lives are actually decided—or a secret college of mystical masters who determine the course of world history—is indeed an attractive concept to those inclined towards paranoia in any of its forms. Many psychologists will tell you that, indeed, a fascination with the occult itself is one possible manifestation of the paranoid-schizophrenic personality (a claim to which the late psychiatrist R. D. Laing would have agreed wholeheartedly, and perhaps enthusiastically!). It is similar to the emotion that motivates study of the "secret government," for instance, or of conspiracies involved in presidential assassinations. Unfortunately for all of us, one's paranoia is generally justified in the latter matters, so why wouldn't they be in the former?

Well, one can *believe* in an assassination conspiracy without the benefit of proof; or one can simply prove it to oneself through studying the literature and taking an objective view. The same is true of the astral plane. One can either simply *believe* in its existence, without having any more proof than some bewilderingly obtuse description in a popular handbook of the supernatural . . . or one can go about attempting to prove it to oneself by actually following the various physical and mental exercises insisted upon by those who claim to have visited there personally.

The end result in both cases is virtually identical: one comes away from the conspiracy literature convinced there is a conspiracy, but not sure what to make of it or who the major players really are. One comes away from attempts to visit the astral plane similarly convinced it exists, but uncertain as to how far such a conviction can be pushed. Fortunately for us, the extent to which we master the astral plane depends solely upon our will to maintain those practices that lead us there in the first place. Unfortunately for the conspiracy theorists, the only way they can be sure of their theories is to descend into the underworld of organized crime and intelligence agencies, right-wing hate groups and left-wing fanatics, and spread around a lot of money and muscle. It is a little easier to penetrate the mysteries of the astral plane . . . but not much.

Just as one needs a "cover" for infiltrating a secret society, one needs an "astral body" to investigate the astral plane.

According to the Hindu mystics, everyone already possesses one of these. It is simply a matter of becoming conscious of it. In this cosmology, even though events are taking place "behind the scenes" on the astral plane, our astral body is already aware of these mysterious machinations. Our conscious self, however, is not. Why make the conscious self aware of unconscious occurences? Why, to affect them, of course.

"Magick is the art and science of causing change to occur in conformity with will."[2] That is one of Aleister Crowley's most important maxims. It is a corollary of the Hermetic axiom: *As Above, So Below,* which was echoed in the New Testament's *What you seal on earth shall be sealed in heaven.* It is the insistence on an interdependency between microcosm and macrocosm, between conscious life and unconscious life, that strikes us as the single most important element of all occultism, all magick, all mysticism. It is the one aspect of magick that separates it from religion. While organized religion teaches that all that happens to us is "the will of God" and should be accepted—albeit with prayer for help—magick teaches that "the will of God" and the individual Will are virtually identical. Why else would Christ give Peter the keys to the kingdom, telling him that whatever he signs on earth will be cosigned by the Creator? What else could the implication be of *As Above, So Below*?

Therefore, although our existing astral body may be aware of the dynamics of this invisible—or unconscious—world, our conscious mind is not. Oddly, however, our unconscious mind is completely aware of our conscious state. It's a one-way street for most of us. "The Will of God."

To rectify that situation, the Indian sages developed practices to make the conscious mind aware of the unconscious processes that were taking place within us. This involved a steady, step-by-step method of gradual opening of the link between the conscious "body" and the astral body, using the Will as the guiding force. Once this link was made, information could travel freely in both directions: impressions received from the astral plane could help us orient our lives in harmony with macrocosmic tides, and our own desires could be transmitted directly as *orders* (rather than simply wishes) to the astral plane, thereby bringing both halves of our existence into greater unity.

Thus, if the reader will permit me to pursue the conspiracy analogy one step further, we have moved from simply *investigating* the conspiracy to becoming a card-carrying *member*. As those in authority used to tell us during the Sixties: you can only hope to effect change through the System. Sure, but which System? The overt one of elected government, or the covert one of secret deals and shady characters? If you're reading this book in the first place, I think I know where you stand on that issue, and you should be comfortable with the secret government metaphor. You must realize, however, that this secret goverment is relatively benign compared to that of a Republican administration!

The Chinese are not as sanguine about the existence of an astral body as are their Hindu siblings. To the Chinese sages, as to the Malekulans, this astral body does not exist unless the initiate *creates* it, consciously, using very much the same *type* of techniques as the Hindu mystics. The basic elements of the astral body were already there within our bodies, but they were not "hooked up." The various essences and effulgences had to be cultivated; the *jing* husbanded tenderly as a green shoot in a spring garden until it could blossom. Otherwise, upon death, the various "astral" elements within our beings would dissipate and the chance for immortality lost forever.

The Chinese did not automatically embrace the concept of reincarnation. To them, this idea that everyone would be born again and again until each individual attained perfection was, well, laughable. There was an afterlife, to be sure—at least, according to Confucius—and if a person were to die without a family to perpetuate his or her memory through the placing of a memorial tablet and the burning of incense on special days throughout the year, then that person would become a hungry ghost, and a ghost to be feared. Otherwise, the deceased remained, quite simply, dead.

Peacefully, perhaps, but dead nonetheless. There was no constant return to earth with the end result being Godhead. Various individuals—such as Confucius—might have attained to a particularly blessed mode of postmortem existence and reign in an equivalent atmosphere as a Christian saint, but the state of *immortality* was one to be attained in this life, while alive. In fact, there were various levels of immortality that could be attained, from the "merely immortal" rank to an immortal in one of the "upper heavens." In any event, the actions that need to be taken to ensure one's position in the afterlife have to be taken *now*. There will be no second chance. As if to underline this concept, the Chinese alchemists and occult philosophers of the early centuries of this era even believed in a form of ritual suicide to accompany the final commitment to immortality.

Honoring one's ancestors was more a tool of enforcing filial duty towards one's family and, by extension, one's emperor, than it was a carefully thought-out cosmological scheme. Heaven, in these cults (and the author includes the Christian heaven, particularly the Roman Catholic variety), is guarded by the State or its equivalent: a hierarchy of humans and some deities who decide who goes in and who stays out. Strict adherence to the rules of the State determine one's fitness for admission to a halfway decent afterlife. This afterlife is modeled after the very human monarchical system of government, with an Emperor at the top and many, many descending ranks of immortal beings below. To those of us who yearn for a more democratic afterlife, this must appear to be a truly disheartening prospect. But, then, after all, can one *elect* God?

Yes, if that God is yourself.

With the importation of Buddhist doctrines into China, and the resulting acceptance of reincarnation, the Chinese sages gradually came to believe that if reincarnation existed, it was an endless cycle: one that had to be broken anyway . . . one that had to be defeated *in this life*.

The Daoist sages—being naturally averse to rules of the State—developed a way around this: create your *own* body of immortality, your own astral body, and go directly to heaven yourself without waiting for permission. Not only that, but go right to the Garden of the Immortals itself: dwell with the likes of Confucius, and Lao Zi, and all the other saints and gods of the Chinese pantheon. Anyone could do it. The Gate is there. You just have to know how to open it.

This individualistic approach to religion has always been condemned by the large organized churches. Direct access to God is frowned upon, denied, or ridiculed. Part of the reason for this, of course, is that those who attempt this dangerous journey on their own quite often appear insane to their neighbors. The journey takes its toll, and those who pursue a solitary charge up the Sacred Mountain to Heaven's Gate are regarded as psychotic, schizophrenic . . . or worse. As heretics, satanists, and witches, these people have been burned, hanged, tortured, and dismembered.

Secret societies have managed to control the fallout from these experiments by structuring the initiatory process in such a way that psychological safeguards are learned at the same time as illumination is received. A society of like-minded fellow initiates can protect the seeker both from the world at large and from him or herself. Warning signs can be recognized early, and appropriate steps taken to ensure that the initiate does not fall off the end of his or her own, inner world and into eternal chaos.

One way they have done this is through the patient creation—or discovery—of the astral body. The astral body is not only the easiest means of observing what goes on "behind the curtain," it is also the best safeguard against psychic dysfunction. It enables the person to live simultaneously in *both* worlds, in *both* modalities, without having to sacrifice one for the other. In other words, it allows the initiate to retain the ego and consciousness relatively intact, while the astral element of the personality takes all the psychic heat. In later stages the ego may experience dissolution—a necessary phase of higher illumination—but by that time the astral body has become strengthened to the point that it helps to "carry" the conscious self through the world without attracting unwanted attention. What we are discussing is a very fine balancing act between the conscious self and the unconscious self, a process that can only be safely accomplished with the formation of an astral body: a *sub*-conscious "ego," if you will, that is the repository of the initiate's core identity and that allows the initiate to penetrate other levels of awareness without sacrificing sanity. It is a means towards a kind of *controlled insanity* that is so necessary to genuine attainment in occult or magickal practice.

Also, the means used to create the astral body are based on the initiate's own, personal cosmology as informed and delineated by the occult or magickal system being used. In other words, a perfectly structured cosmological system—whether it be the Qabala, Wicca, Christianity, Hinduism, Thelema, whatever—has, by definition, built-in safeguards and protective mythologies that will assist the initiate at critical moments of the quest. A cosmology is more than just a pretty picture of the World: it is also a *process*; it is the road map and the road, at once. Once this cosmology is carefully committed to memory—i.e., planted in the unconscious mind—one finds the way clearly marked.

Of course, one is aided by the fact that the system chosen is one that is calendrically *current*: i.e., in tune with the astral forces currently in place. As the Old Age dies—the Age of Osiris, of Pisces—and the New Age is born—of Horus, of Aquarius—we live at the midnight hour between the two ages, the time of "rising yang," the time of the most potent rituals, when occult power—*Heka*—is being born in the cosmos. The Goddess of the Crossroads and Goddess of Occult Power—*Hekaté*—is there to guide us, to bring light down into the Underworld before us as we begin our descent to the source of Yang, as we begin the Pace of Yü, the Ceremony of the Walking.

* * *

I have spent so much time in discussing governments and organized religions in this chapter for a purpose: the formation of an astral body is a political act, an act of social division. It is an act of personal responsibility and commitment, the result of long periods of intense—often solitary—practice. Once the astral body is successfully formed, the individual is beyond the normal control mechanisms of either government or religion; i.e., the individual is incapable of being "brainwashed" or of being subject to any of the various subtle forms of mind control that exist in the media, advertising, etc. That is because these techniques normally bypass the consciousness of a human being and invade the unconscious directly, through subliminal methods or by identifying the goals of the church/state with unconscious goals through the manipulation of, for instance, the libido. The astral body, however, is a deliberately formed *sub*-conscious mechanism and as such guards the entrance to the unconscious mind like a watchdog, like Anubis at the Gates of the Under- or Otherworld. It increases conscious sensitivity to subliminal messages and deprives them of their psychic attraction. It enables the individual to make more intelligent choices and to interpret information based on actual content rather than form.

Essentially, the individual becomes a priest/priestess in his or her own religion. It is an act of loyalty to one's Self, but an act of rebellion against the Realm. By forming an astral body, one obtains the keys to the Gates and does not have to wait for St. Peter to be in a good mood, or to look the other way. One comes and goes at one's Will; one peeks behind the curtain to see of what the Wizard is really made.

The Formation Process

We are using the term "formation" rather than "creation" so that we can incorporate elements of both the Chinese and the Indian techniques without getting entangled in theological discussions as to whether the astral body already exists or has to be created. In either case, the authorities agree that the astral body has to be "awakened" or somehow "empowered," to use a contemporary cliché. For our purposes, the term "formation" should suffice.

The *Necronomicon* is clear that a period of purification should precede the Walking, and goes so far as to specify a lunar cycle in combination with abstention from sexual intercourse, with the sole exception that one may "worship at the altar of Ishtar save one does not lose one's essence." This latter is, of course, analagous to the Daoist strictures concerning preservation and cultivation of the *jing*. Worshipping at the altar of Ishtar is probably equivalent to the *ou-jing* practice of dual cultivation, either with the "Celestial Maiden" or with a human partner; the pronouncement against losing one's "essence" is obviously a warning to the male not to ejaculate, a sexual practice known in the West as "karezza" and treated at length by various other authors.

The "calendar" insisted upon in the *Necronomicon* is lunar, and prescribes the space of a lunar month between each Gate. Also, there is a lunar month of purification that takes place before the first Gate. Hence, this gives us a total of seven lunar months for the entire practice.

The first month is crucial, for it establishes the pattern of "Walking": sexual abstinence, fasting, a three-day "Black Fast" of water immediately preceding the ritual, etc. These practices are designed to enable the operator to develop an astral body capable of making the voyage to which he or she is committed. A month is not a long time for forming an astral body; in fact, to some Teachers it would seem ridiculously short. However, the entire process of passing through the Gates as outlined in the *Necronomicon* has the effect of slowly building up the astral body over the entire seven-month period, very similar to the way in which the deceased passes through the seven Underworld Gates in the Egyptian mummification process.

The very nature of the *Necronomicon* rituals themselves

accelerates the formation of the astral body, as the material is so ancient and so direct, without a lot of the frills and complex instructions of the later cults, that one is left to figure out the method largely on one's own. One falls back on the unconscious, which, operating in tandem with the mythic material of the *Necronomicon* and triggered by the ancient words of power that have not been used in over six thousand years—except among extremely secretive cults in Mesopotamia—helps the operator to break through layers of resistance from the ground up. From the deepest layers up to the conscious layers. Most modern cults employ rituals based on those of antiquity, but within a cosmological framework that is the invention of the later Qabalists or Theosophists; that is, a schema is imposed from the outside in. From the conscious layers down.

In the case of the *Necronomicon*, however, this process seems to be reversed. The God-forms and words of power—*Hekau*—are those that were the *first* names and *first* words of power in recorded history for attempting the Otherworld journey. As such, they are part of a virtual genetic legacy common to all of us of whatever race or religion. If the methods of the Sumerian magi were known to people as far away as the Chinese, the Malekulans, and the Mexicans, we can suppose that the rituals of the *Necronomicon* are representative of the original Cult of the Stars. There is a memory encoded within the rites and the words of power that is triggered once the book is opened, once the conscious decision has been made to open the Gates. This memory begins to work its way upward from the deepest layers of the unconscious mind to meet the light of day in the consciousness, and the astral body forms itself around that memory, clothing it in the material of the stars. It is a moment of recognition, as the Yang Body rises from the Underworld at the hour of midnight to meet the Yin Body, and to form a vehicle of tremendous magickal power.

The *Necronomicon* has been mentioned in various journalistic accounts of crazed teens performing "satanic" rituals in the woods in the middle of the night, carving up animals or each other. This is the fallout from the power of the Book when it is used without the necessary purification periods. The term "purification" should not be understood in some modern Christian moral sense. It is a technical process that is required before any tampering with unconscious material can take place. The effects of the *Necronomicon* rituals are potentially explosive. Those undertaking to open the Gates without forming the astral body according to the instructions are tampering with their own sanity in such a way that they will never come back again, will never regain "consciousness." As the *Magan Text* amply illustrates, when Inanna returned from her Descent into the Underworld, "the demons preceded her."

The methods of the Chinese and Hindu mystics—or equally of the Golden Dawn and A..A.. adepts—can be used to form the astral body if further guarantees are required by any individual who decides to undertake this Quest. It should be pointed out, however, that one should not deviate from the instructions as set down in the *Necronomicon*, however much one may wish to amplify them with rituals of one's own selection or devising. The altar faces the North, the place of the Underworld, sacred to Set in the Egyptian mysteries and to the midnight sun, and the direction in which the Shangqing initiate faces when he approaches the infinite. On that altar must appear the statues of the God and Goddess: the polar opposites, the North and the South of the Egyptian mysteries, the twin poles of the sacred lodestone, the "dual energies" of the Chinese magicians. Care must be taken to invoke *both* of these divinities, to perform "dual cultivation."

The Watcher must be invoked to take its place as guardian of the circle. There are four lamps in the four cardinal directions, and the incense appropriate to the ceremony is burning in its brazier after the invocation of the God of Fire. The Three-in-One—ANU, ENLIL, and ENKI—must be invoked.

And, upon the ground, is inscribed the Gate that one will enter during the ceremony. Like the Chinese silk cloth on which is embroidered the constellation of the Northern Ladle, the Gate is both the Map and the Road itself. It must be drawn clearly and accurately.

By the time the midnight hour has arrived, the sensation of being about to embark on an extraordinary journey will have become nearly unbearable. The accumulated tension and expectation of the past month will have its effect on this night. Perhaps, the conscious mind will have been fortified by intensive research into the literature of the Gates, in mythology, religion, psychology, philosophy, even the sciences. Ideally, the incantation appropriate to the Gate will have been memorized beforehand, although a written copy of it *must* accompany the operator during the ceremony, for the mind can play strange tricks. If the incantation is memorized, then the author recommends that it not be recited aloud prior to the ceremony itself; that is, it should be memorized silently without verbalizing any of the words of power.

The area selected for the ritual should conform as much as possible to the instructions found in the Book. Under the open sky is best; otherwise, a room must be selected comparable to the chamber in which the Shangqing mystics gathered, which is comparable to the underground vaults in which the Egyptian mummification process took place. There should be nothing suspended from the ceiling. The reason for this will become obvious to anyone who perseveres in journeying through the Gates.

Clothing should be clean and simple. There are designs for several articles of ritual apparel in the *Necronomicon*.

The Seal of the Spirit of the Gate is at hand, and the act of whispering its name upon it signals the formal part of the operation. After that, the operator begins the circumambulation—the Walking, or the Pace—beginning in the North and continuing clockwise around the periphery of the Gate, as inscribed on the earth, the same number of times as the sacred number of the Gate itself. After this, the operator falls upon the ground—prostrate—and looks neither to the right nor the left: i.e., must be single-pointed in goal as well as in vision. There may be sounds and sights beyond the periphery of the Gate, but *these are to be ignored.* One cannot stress this too much.

According to the rubric, one will eventually see the Gate opening "in the air above the altar." The implication is that one is prostrate on one's back, so that a line of vision from the ground to the air above the altar is not impaired. The other possibility is that one is indeed prostrate on one's stomach, but that the head is bent upwards at an angle so the altar can be clearly seen.

Although both positions work, the former is more comfortable. The spinal column is still straight along the north-south axis in either case, and this seems to be important from the standpoint of both the *Necronomicon* and of the other mystical techniques we have discussed. That the altar should be *faced* is what is important, so that the vision is trained on the Northern Quadrant and in no other Quadrant.

After this, the *Necronomicon* is silent, except to say that the Gate will be opened and a Name—a word of power—communicated that will provide the operator with a kind of password to open that Gate in the future. This is followed by a telltale sentence:

> When the First Gate has been entered and the Name received, thou wilt fall back to Earth amid thine Temple.[3]

This concept of "falling back to Earth" agrees completely with what we have already said about the Pace of Yü in this context: an essential aspect of oneself has left the Earth as we understand it. The Sword of the Watcher is left to guard the physical body against attack by forces terrestrial or other.

(To those readers who have understandably doubted the

authenticity of the *Necronomicon* over the years, the author humbly submits the following in its defense:

(The works describing the Pace of Yü and the other Mao Shan techniques of Shangqing Daoism were not available in English before 1977, which is when the *Necronomicon* was first published. The reference volumes the author has used to clarify the mythos of the Gates are all dated years after the first appearance of the *Necronomicon* on booksellers' shelves. The possibility that this entire ritual scenario could have developed independently by the editor of the *Necronomicon* without prior and in-depth knowledge of the Chinese texts strains the limits of coincidence. The only plausible explanation is that the *Necronomicon* is a genuine survival of ancient practices that *predate* the secret Chinese texts, perhaps by as many as thousands of years.)

Once past the MARDUK Gate, it is possible to descend into the Pit itself—past the Gate called GANZIR—and to hallow the Underworld. Marduk, as vanquisher of Tiamat—the Sumerian Leviathan—gives the power necessary to preserve oneself in the face of the terrifying sights of this, the deepest layer of the human mind and brain. Within the Pit dwells the Minotaur, the real essence of the Slaughtered Bull, the bloody, slime-encrusted amniotic sac of the *ka*. The experience is different for every individual . . . and eerily the same. One tempts grave psychoses, demonic possession, and loss of super ego when one delves this far below the level of conscious thought. For that reason, it is not to be attempted before the MARDUK Gate is negotiated successfully. That is, not before the astral body has been formed to such an extent that it is able to withstand the psychic shocks that will surely come when any attempt is seriously undertaken to open that particular Gate.

A careful reader will have noticed by now that the terms Otherworld and Underworld seem to be used interchangeably. To the ancients, indeed, the Underworld was not solely a place of death and destruction. We who are inheritors of the Judeo-Christian tradition have been trained to believe that the

Underworld is just another name for Hell, a place under the control of Satan, where the damned souls go to spend eternity in everlasting torment. If such a place were to exist—and I'm sure it does in the minds of certain of our more psychotic political and religious leaders—it would be only a corner of the Underworld we refer to within these pages. For the Underworld is just another name for the world that exists beyond the reaches of the realm, a world that exists in the stars. Going there *is* perilous; everyone has their personal demons, and the Underworld is where they live.

Amidst the silence, solitude, and darkness of the preliminary Gates, faces rise up to greet us from we know not where, to threaten us with unimaginable terrors. Anyone who has spent some time listening to the *Magic Flute* and paying attention to its scenarios will realize that the fearsome quality of the Underworld is directly proportional to the lack of purification in an individual's soul. There is no way to glide through the Seven Gates in an effortless way. There will be pain and the gnashing of teeth. The shamanic quest can promise nothing less than that, for if it did, it would reward the initiate with nothing more than a hearty handshake and a funny hat, like the lodge brothers of so many Fifties sitcoms. Those of us who are drawn to the Quest are those of us who are propelled by "personal demons," and they must be met and neutralized.

For, after the personal demons have been mastered, other Demons await: the Gods of the Other Side.

The Ancient Ones.

CHAPTER 6: The Primordial Conflict

It is often lamented that the lot of humanity seems to be endless strife and constant warfare. Christianity is said to be a "religion of peace," yet the language of war is to be found in every Gospel. Our earliest recorded faiths describe the origins of the world in terms of a cosmic battle taking place between two sets of Gods: a Serpentlike God or group of Gods that dwells "below the earth" or "in the abyss" or "in darkness," and another God or group of Gods that comes from the stars, or are somehow identified with specific stars or planets.

Thus, Marduk engaged his "mother," Tiamat the Leviathan, in a battle for supremacy over the cosmos. This battle took place long before humanity was created, for we humans were made from the flesh of the slaughtered Serpent and the breath of the victorious Marduk.

This same myth is repeated in the Egyptian and Aztec mysteries. The Primordial Ocean is the source of all matter, but the God who defeats the Serpent of the Ocean in combat then goes on to create humanity out of the remains of the slaughtered serpent and with some of his own, divine "essence."

We can easily sense a primitive explanation of Dr. Crick's "directed panspermia" idea. The "essence" that came from the "Gods" in the "Stars" would be, of course, the genetic code: a thing not visible to human eyes, an incomprehensible message from our point of origin, something that can only be explained by means of allegories. By preserving one's *jing*—one's sexual fluids—one is naturally preserving one's genetic code as well, at least in terms of chromosomes (for the genetic code is available in its entirety in every cell of our bodies, something that was unconsciously understood by every *voudoun* priest or priestess who ever cast a spell with a victim's blood).

The stuff of the Primordial Serpent would correspond to the actual *physical* material of our bodies: the "organic" elements, as Freud might have said. These are the component elements of the flesh that are organized by the code: carbon, oxygen, hydrogen, etc. As many others have pointed out when discussing the astonishing complexity of the genetic code, although all the elements necessary for the code exist on earth, the chances of the code somehow "evolving" naturally are so astronomically high as to be all but impossible. The proposal that the code was somehow "seeded" here on Earth from some extraterrestrial source becomes, oddly enough, more plausible than the evolution theory. (This gives new meaning to the term "astronomically high"!) This is even more plausible now that humanity has already ventured off the planet and walked on the Moon. If we can do it, why not the residents of some other planet revolving around some other star?

Also, we represent the genetic code by lengthy strings of letters, combinations of four basic "elements" that correspond nicely—in number—to the four "Platonic" elements that are the basis of so much occult practice. Even in Asia, where five elements are recognized, the extra element is seen as the element of the "center," the other four corresponding to the four quarters, four winds, etc. Diagrams of four equal sides are important to Jungian analysis as symbols of psychic wholeness; such as these are the famous mandalas of Indian and Tibetan Buddhism.

The genetic code is a thing of numbers and letters, of chemicals and heredity, of memory; it is a book in which we will one day be able to read the story of our origins and write the future with deliberate intention. Perhaps it is within the pages of the genetic book that we will one day discover the lost Word of the Masons, the Logos, the unpronounceable name of God. This is the breath of Marduk, the divine essence of all the star gods in every culture that has contributed to the Origin of the human race. As the self-professed atheist Paul Davies says in his book *The Mind of God*, we are "children of the universe—animated stardust . . ."[1] Even the scientists have come to the realization—albeit grudgingly—that "Every man and every woman is a star" (Liber AL, I:3).

It has become a cliché to speak of the Earth as a living being in its own right; a noble conceit perpetrated by movements organized to protect the ecology and bring an end to pollution. To those organizers, humanity is seen as a kind of virus, reproducing like mad and destroying the host organism like so many cancer cells. Indeed, the only potentially immortal cells on this planet are the cancer cells and the sex cells. Each reproduces frantically as long as the host organism is alive to support the growth. They can be seen as the microscopic analogues to our Manichaean adversaries, Light and Darkness, Good and Evil.

To go even further, this virus—humanity—is now sending its cells all over space in an attempt to infect other organisms wherever they can be found as the host organism, the planet Earth, is showing signs of dying.

Certainly, a case can be made that animals, for instance, do not pollute or otherwise destroy their environment. Compared to us, they *are* the environment. These "lower" forms of life do not feel the need to use a system of writing, to build tools, or to get off the host planet. They don't smelt ores or watch television. They don't have books or indoor plumbing. Indeed, from one perspective, humans are definitely an alien race. We are uncomfortable with this planet—with its geography, its weather, its seasons—and we fight tremendous wars over trifles like gold and whose God is "real." We yearn heavenwards, towards the stars, desperate to leave the hellhole we have created around us, fearful of our fellow humans and their intentions on our homes, our livelihoods, our very persons. We even have to make rules governing *sex*, something the animals would certainly not understand.

Then what the hell are we doing here?

The existentialists write of our state of desperate unhappiness here on earth, and of our vain hope for a better life after death. If God is truly dead, then we are all on our own. (*Spirit of the Sky, Remember. Spirit of the Earth, Remember.*) It's time to abandon the churches and start building space shuttles. It's time to stop praying for the stars and time to start *going* there.

If the Gods *have* forgotten us, it's about time to show up on their doorstep and *remind* them.

But before we do—before we rush off and venture into deep space where angels fear to tread—we had better review all those old myths one more time. The warnings are there, if we care to heed them.

The Myth of Origin

Whether it was from the blood of one and the flesh of another, or the breath of one and the flesh of the other, humanity is nothing more than an artificial creation composed of the elements of two, preexisting forms of existence: an "upper" God and a "lower" Monster. The Zoroastrians, Manichaeans, and other Middle Eastern gnostic-type sects would picture this as "spirit descending into flesh"; for many of these cults, procreation was evil because it increased the amount of flesh or matter on earth, which was of the Devil (or Monster, or Demiurge). This idea that flesh is Evil and spirit is Good is the main propaganda effort of one side in this conflict. This antimaterialist stance is reflected in other belief systems, as well, including the horror the early Christians had for such pagan figures as Pan, a Nature God.

This theology was shared by such groups as the Cathars—the "Pure"—where a form of ritual suicide was permitted, à la the Chinese Daoists. After all, when flesh and matter are evil, then it must be the height of good taste to destroy them.

And, if flesh is evil and procreation is counterproductive, then obviously sexuality is sinful. The sexual urge is a function of the flesh—according to this theory—and not of the spirit. The spirit becomes "imprisoned" in flesh when a baby is born. The sex act is therefore a magickal invocation of spirit, forcing spirit to descend into matter. Naturally, the imbuing of inanimate objects (matter or flesh) with magickal power (spirit) is one of the chief aims of magicians the world over. It is also the core rite of the Catholic Mass and the Eastern Orthodox Divine Liturgy, encoded in the *Epiklesis*, a secret prayer recited by the priest which invokes the Holy Spirit into the bread and wine to effect the transubstantiation of those materials into the body and blood of Christ. In the Catholic Church the sexual act—missionary position, please—is permitted *only* when there is a chance of procreation. The only birth control method endorsed by the Church is the famous rhythm method, and it is always astonishing to see how many Catholics cannot keep a beat.

Of course, the Catholics would much prefer the voluntary celibacy of all communicants. Their priests and nuns are, by canon law, celibate and may not procreate. The war against sex by the Church—particularly against sexual pleasure—found itself in an uneasy truce by agreeing to the sexual act only between partners married by the Church and only for the purposes of having children. It was a political judgment call rather than a theological pronouncement: if you were going to have sex, then at least produce more dues-paying Catholics, please. Christ, in the Gospels, says of heaven that there is no marrying there. The implication is, of course, no sex. But in the Platonic worldview, repeated as recently as the *Communist Manifesto*, the Utopian ideal is a nation where marriage is a thing of the past and all adults are the parents of all children. It can be argued that the potential for child abuse in such a society where all adults are responsible for the welfare of all children would dwindle.

This quasi-Cartesian dichotomy between "flesh" and "spirit"—like that between mind and body, or blood and brain—is a survival of these very old beliefs concerning the creation of the human race. This is not to be confused with the creation of the cosmos, which is usually a seperate act in the creation myths. Humans were evidently not created first and allowed to float around in space before the worlds were made; for some reason this version of a creation myth is found nowhere, leading us to appreciate the perception of our "ignorant and unlettered" forebears. The world was made first—in the Egyptian myth, via the masturbation of the first God—and humans came much later, only after the battle in which a dreaded Monster is slain and carved up to make the planets, the stars, the sky, and our own Earth.

Our divine forebears, as has been mentioned many times already, seem to have come from a star or cluster of stars in the Northern Dipper. In fact, there is evidence to show that our own Star, the Sun, is actually *part* of this star cluster, part of the Dipper itself.[2] However, there is also a tradition that cites the constellation Draco as the origin. If we were to apply a little archaeoastronomy to this concept, we might say that the slaying of the Monster is really a reference to the precession of the equinoxes, when Draco ceased to be the circumpolar constellation *par excellence* and was gradually replaced by the Dipper, some five thousand years ago. This could be the origin of the idea that a "serpent was slain." In that case, the whole myth is nothing more than an attempt to describe something as mundane as the precession of the equinoxes.

Except, when the first pyramids were built, Draco was still the center of the northern sky and by this time the creation myth was already old. The Sumerians also had this particular creation myth, when Draco was high and bright in the northern sky. The identification of Draco with the Ancient One, Tiamat, is therefore not tenable. Leviathan is not merely a disguise for an astronomical phenomenon; it represents some other reality.

A Fairy Tale

Imagine this, if you will.

Let us assume that the "underworld" refers to a place. Let us assume that this place is "under" the world, i.e., that it is somehow below the Earth, the planet Earth. In other words, it could represent a star or constellation that was once visible to the denizens of the middle latitudes but is no longer. For instance, Sirius disappears below the Middle-Eastern horizon for a certain number of days every year. What if another constellation has disappeared *entirely*, never to reappear? The nighttime sky of the southern latitudes is completely different from that of the northern latitudes.

And what if, among the stars of this constellation, there was a planet from which living beings came to our planet one day. These beings may have been somehow loathesome, unsavory, cruel, or vicious in ways we can only imagine. The Ancient Ones. Perhaps, when they appeared on Earth, their star was above the horizon: a visible source of this demonic life-form. Later, the star began to sink until it was visible no longer.

That these creatures were unpleasant was something known to other inhabitants of our galaxy. Once it was realized that these monstrous beings were making trips to Earth, the others decided that something had to be done to stop them. An army was sent to Earth—the Elder Gods—and a battle raged between the forces of Leviathan and those of, well, Marduk. Marduk won the contest, destroying the horde of evil beings, but only after a lengthy fight in which many soldiers from both sides were slain.

Their blood spilled on the earth.

There was oxygen on this planet, and carbon, and hydrogen, and nitrogen, and all sorts of elements necessary for life. But there were no living things as yet. No animals, no plants, no people.

No genetic code.

But now, with the vast quantities of spilled blood and decaying flesh littered over the planet, the basic building blocks of life were suddenly available. The battlefields, once redolent with death and horror, are now suppurating with life. The planet Earth has been "seeded" with the flesh of the Monster and the spirit of the Gods.

Later—perhaps millions of years later—beings arrive to check up on the status of the planet. (Ancient Ones or Elder Gods?) They find people. They try to teach them something about their origins, but the people are still hopelessly uncivilized and lacking in even rudimentary science, mathematics, language, astronomy, agriculture . . . even after having populated this planet for *hundreds* of *thousands* of years! How could these ugly, defective, deformed beings—half Monster, half God—hope to understand anything of their stellar origins?

So the story is told in its basic form. The star from which the Bad Guys came is no longer visible in the middle latitudes. You can't point to it in the midnight sky. It is "below the Earth." The people of Earth do not understand this at all. How can a star be below the Earth? The star people sigh. "One day," they say (like frustrated parents), "when you're older, you will understand. But be wary, for we have not heard the last of those foul beings from the hidden star."

This would have filled the people of the Earth with terror, for they could not *see* the star that was the source of all their unhappiness. It was "under the world," hidden, like some malevolent monster in the ocean. You could never tell when it would strike. You would never even know when it was there.

But *which are the Monsters and which are the Gods?*

At some point someone must have doubted the version received from whatever Beings happened to dispense it, and began to find out where the truth was hidden. Using techniques similar to those discussed within these pages and in the occult arts of all Earthly cultures—remnants of some sort of "star travel" methodology left behind by our extraterrestrial visitors—these skeptics began the voyages on their own, their goal being the discovery of our stellar origins. Someone, at some point, said, "Why take *their* word for it? How do we know that the God of today was not the Monster of yesterday? Or, more to the point, who needs *either* of these races? We are humans! The dispossessed orphans of the universe. A plague on both their planets!"

The above scenario is not meant to be history, only a suggestion. While the techniques employed by the magicians and shamans are valid methods for probing the depths of the Underworld, for attaining Unity (what Jung called "individuation"), and for developing paranormal abilities and insights, there is no need to swallow the propaganda along with the powers. We are in the position of a bunch of impoverished dirt farmers being handed fistfuls of cash by some seedy politician who urges us to vote for him or her even though the Party's platform is repugnant to us. We can use the cash. We take it. We don't have to give the politician our vote. (If you don't believe that, then you believe that the politician is right to buy our votes with cash.)

We don't have to believe that Evil is Good and Good, Evil, just to taste a little immortality. We don't have to believe in everlasting damnation for heretics, Lutherans, and witches in order to be saved. We don't have to despise the infidel, or the Jew, or the Catholic in order to be guaranteed a place in Heaven. Because "Heaven" *means* "the heavens"—the stars—and each of us are already guaranteed a place among the stars if we are brave enough to make the trip.

No one can prevent us from taking the steps we know exist to some form of immortality. Knowing who we are and where we came from is our right as humans. Conquering death—in some form, in some way—is our birthright. We don't have to accept the arcane moral laws and biases that are stuffed down our throats by the vested interests, the Gods who have abandoned us here on this planet so long ago and who *now* may have some use for us, some place reserved for us in some battle plan of which we have no understanding, no hope of comprehension.

Our responsibility is to our own survival, both here on Earth and after death. Whoever or whatever helps us in that Quest can be useful to us at that time, no strings attached. Whoever or whatever stands in our way of achieving this simple goal of survival is anathema. It is not the destiny of the human race—the race on Earth—to serve as cannon fodder in some real-life version of *Star Wars*. The churches have failed us, and so have the governments. Our souls have become our own responsibility; they always were. Only a very small handful of people realized that, and they were persecuted by Church and State.

And still are.

God is the only safe thing to be.

The combat in which we engage when we commit ourselves to the Quest will be painful and will threaten our sanity, and the peace of those around us. We will confront demons. Loved ones may abandon us. Desolation will be our neighborhood, at least for a little while. And when we have beaten these demonic forces and subdued the Enemy, "conquering death by death," and are reborn, become Osiris, the battle has not ended. For the interior battle we wage for the dominion of our own souls continues in the exterior war for *all* souls, for the survival of our Race, the Human Race. As above, so below.

Do not think that the Race to which we refer is white, or black, or yellow, or any other color. This warfare between races according to colors and religions and languages is a device to keep the entire Race disunited, the easier to conquer from Outside. It is a cynical device of the Ancient Ones. As we have seen, there is a root-religion to all religion, a root-magick to all magick, and our Race is likewise One and not Many. The destruction of the great library at Alexandria . . . the Holocaust . . . the Cambodian genocide . . . the Klan . . . the burning of books and the burning of people—are these not all different manifestations of the same suicidal impulse?

The Myth of Combat

The Gods—those to whose Heaven we all aspire—seem to be in eternal conflict with each other. Marduk and Tiamat, Horus and Set, Tezcatlipoca and Quetzalcoatl, Zeus and everybody. The Other/Underworld is a busy place.

In ancient Egypt it was not unusual to worship Set. There were temples to Set all over Upper Egypt; he was an important god until the Osiris cult became popular and Set was relegated to a kind of second-class status. Set, we should remember, is a God of the Dipper, as is Tezcatlipoca; these are the "bad guys" of the Egyptian and Aztec religions, respectively. Yet, they are identified with our Place of Origin. They are clever, intelligent, and highly sexed. Like the Kennedys. For some reason, that makes them evil.

Horus and Quetzalcoatl, on the other hand, are relatively innocent. Osiris is so passive that modern authors refer to Osiris more as a *process* than an actual being. Although both Osiris and Quetzalcoatl are potent gods, represented at times by phallic monuments and stories of great sexual prowess, their sexuality is often only the socially acceptable kind. When Set sodomizes Horus, Horus resorts to a ruse (aided by Isis, to whom he runs like a child) to convince the Elder Gods that he, in fact, had sodomized Set. It would never occur to Horus to sodomize anyone, but Set knew what humiliation Horus would bear if it were known that he had been sodomized. There was no onus in being the *sodomizer*, only in being the *sodomizee*. This entire episode is so full of complex social significance that a detailed study is beyond the scope of this book. Suffice it to say, however, that of the two, Set is the more sexually experienced and adventurous. As is Tezcatlipoca—the Aztec Northern Dipper—when compared to Quetzalcoatl, the Plumed Serpent.

Set—after being trounced by Horus—is generally engaged in defeating Apophis, the Serpent, and preventing Apophis from destroying the Solar Barque. Tezcatlipoca is engaged in fighting the Plumed Serpent, Quetzalcoatl. So, here we have parallel instances of Northern Dipper deities fighting cosmic serpents. In the case of Tezcatlipoca, however, the Serpent is the *good guy*. In the case of Set and Apophis, it is acknowledged that one day the Serpent will rise and consume the Worlds and bring creation to an end. A *foreordained* and *predicted* end.

Just as most ancient cultures around the world had a Flood legend in their store of mythology, so they have a cosmic warfare legend and a serpent legend. Even Ireland, which doesn't have any serpents, needed a serpent myth so badly that they had to invent a time and a saint—Patrick—for a myth of driving the serpents *out*.

The term "serpent" has been shown to occur in some mystery religions as a code word for "initiate," particularly when the serpent is depicted in ikonography or in texts as having legs.[3] The identification of the Serpent with Wisdom is as old as the Old Testament, and probably much older. The Serpent figures in much occult lore and legend, including the crucified serpent of Masonic and Alchemical allegory. The Serpent, of course, is also Kundalini: a symbol once again of occult power, this time as encoded in the body and subject to being raised by the practices of a specific form of yoga. The image of a serpent nailed to a stake or a tree may simply be another way of expressing Kundalini when raised along the sushumna (spinal column); i.e., a form of realized power: of *Heka*, of which the Egyptian *Ur-Hekau* was such an apt depiction.

In China, where Dragons are considered *lucky* and *beneficial*—as opposed to the Western view that a St.

George or a St. Michael or a Marduk is required to slaughter it—the Northern Dipper deities are seen as the good guys. They are consciously invoked, and by the best and brightest minds in the China of that age. This would seem to indicate that there is some type of equivalence between reverence for the Dragon/Serpent and reverence for the Northern Dipper, the Origin.

When religions became Solar in nature, the nighttime skies were suddenly the source of evil. The Dipper—which is visible, of course, only at night (except during an eclipse of the Sun)—became the domain of evil spirits. Set and Tezcatlipoca were suddenly out of favor, the way Pan and Ra and Cernunnos became the Satans of Christianity. The pagan King Arthur had to be slain to make way for the new Christian kings, and Merlin—Druid of the old Faith—faded from view, imprisoned in a tree by a spell, like Osiris.

Of course, a Solar faith was a lot easier to manage for minds unencumbered with intelligence, for bodies unencumbered with spontaneous sexuality. The sun rises and sets, rises and sets, with intimidating—peristaltic—regularity. If you live your life by a solar calendar and it's off by a few days after a while, just add some days and have a party. (A socially acceptable party.) No need to get *anal* about it. The complex—and highly accurate—calendars of the Aztecs and Mayans, on the other hand, required a thorough knowledge of the stars and the movements of the planets, and record-keeping going back for millennia.

This was too much like work, and was probably an invention of the Devil, anyway. Look to the Sun—as if you could ignore it—and forget the stars of the night. At night, the witches and demons and evil spirits are about, causing mayhem, getting laid, spooking the cattle, and spoiling the milk.

The preceding is designed to show that there is a great deal of ambiguity concerning the morality of either side in the cosmic battle. The Serpents are the Bad Guys in Christianity and Judaism. They're the Good Guys in some forms of

Egyptian and Aztec religions, although they were the Bad Guys in other forms of those religions. Where the religions of the world are in basic agreement that humanity was created after a war in which a Serpent was bested by an Astral Being, there is some doubt as to just who deserved to win. The victory over the Serpent was therefore not a *moral* victory at all; it was simply a military victory. One side just happened to win; no moral point intended. This is particularly clear in the oldest—Sumerian—version of the myth, where the young Gods symbolized by Marduk made so much noise that their parents couldn't sleep, so the latter tried to kill the former. The Sumerians praised Marduk because he won the battle, not because he was morally superior.

That the Serpents are in possession of some ancient Wisdom is generally agreed among all the cults we've studied. But in China the Serpent is aligned with the forces from the Dipper. In Mexico and Egypt the Serpent is destroyed by forces from the Dipper.

Sound confusing?

In American history books today, the United States is identified with Vietnam during the war that bears its name. In Vietnamese history books today, the United States was defeated by Vietnam. It's simply a matter of perspective, not of actual *history*. And as the United States develops trade relations with Vietnam and becomes once more a part of Vietnam's cultural and economic life, then the Vietnamese will say that they have "tamed the Dragon": made the enormous might and power of the United States work for them instead of against them.

The United States, of course, will say that in the end they achieved economic and political victory in Vietnam where military victory was denied to them.

Both are right. How much easier everything would have been had both sides started right away on cooperation instead of aggression.

The same may be true of the Cosmic Conflict. The powers of the Serpent were harnassed by the victors, who could not absolutely destroy the Enemy. Neither side was "right" in any kind of moral sense. The "Thigh of Set" is used in tandem with the "Eye of Horus" to awaken the deceased, to cause the rebirth of the *ka*. Both are adzes, a form of the Dipper. The *Ur-Hekau* is a combination of Underworld serpents and Solar ram. To achieve immortality, both aspects of power are necessary. The Serpents are not evil, neither are the Gods. That doesn't mean, however, that we should trust either side.

As creatures who are unique in that we possess both sets of attributes—Serpent and God—we should labor mightily to preserve this uniqueness. To do otherwise would be to achieve racial suicide. If *either* side should try to destroy what they have created here on Earth, then they should be resisted with every weapon at our disposal. If *either* side tries to bribe us with knowledge or power, we should take what we can use and abandon the rest. There is no need to become converts to a new religion or to a new set of Gods or Serpents, unless it is the only way to guarantee our survival and to promote our evolution as a species. Immortality—like Intelligence, like Beauty, like Strength, like Life itself—is nondenominational.

Evolution, not Conversion. Immortality, not Death for a Good Cause. Once we have defeated Death and discovered the path to Immortality—that is, once we are all truly *free*—then we can look around and ally ourselves with whichever of the Powers we deem sensible and profitable. Until then, however, we should keep our options open.

In the Eternal Combat, Immortality is only the first step in choosing sides.

PART TWO

The Rituals of the Gates

CHAPTER 7: Preliminary Rituals

God and Goddess

Once the place and time for the beginning of the Entrance and the Walking are determined, the preliminary rituals must be performed in order to set the stage for what will follow. These are prophylactic measures designed to preserve both the physical and the psychic integrity of the operator, a person we shall begin to identify as the Magician while realizing that this person may be male or female, straight or gay, without in any way amending or jeopardizing the outcome of the experiment. This is because, crucial to any undertaking involving the *Necronomicon*, the God and Goddess must both be faithfully represented on the altar. The insistence on polarity—here conceived as a psychosexual polarity—is obvious in the *Necronomicon* and is consistent with similar practices in other cultures. Contrary to popular, academic misconceptions, the shamanistic rite is not restricted to members of one sex. Ideally, the gender of the magician may be blurred and confusing to outsiders. This is not the place to discuss such themes; perhaps they will form the subject of a later study.

God and Goddess must be present in some form. Naturally, to the Mad Arab, these would have taken the form of idols. Remember that the Mad Arab is living in a violently Islamic world at the time he is writing: the Muslim armies have advanced as far as Spain and are probing France. The Alhambra is already a gem of architecture. The Holy Land is under the rule of Islamic warlords.

The Mad Arab, on the other hand, is scribbling furiously a tome that deals with the evocation of dark forces and the worship of a God and a Goddess in flagrant violation of the first axiom of Islamic law: there is no God but Allah. Truly, he was mad; but that has always been a prerequisite for shamanic initiation. For the Mad Arab, to insist on the rites and ceremonies we find in the *Necronomicon* is to find himself a Muslim version of a Tantricist: performing rituals whose very nature openly defy the laws and taboos of the society in which he lives.

For us, perhaps, worshipping a God and Goddess is no large abomination in this day and age; the fact that we are applying ourselves to the *Necronomicon* in that context, however, is certainly a dangerous act, considering the opprobrium that book enjoys among police departments and fundamentalist religious sects.

There is no specification as to which God or Goddess is suitable. Any pair of deities with whom one feels close affinity is probably best. They should represent a combination of Power and Spirituality, your noblest aspirations combined with your ideal of greatest spiritual strength.

The Fire God

As in Zoroastrianism and other Middle Eastern cults, there was a particular set of rites and beliefs peculiar to the element of Fire. Fire was at once a rapacious consumer of whatever fuel was available—including wood, homes, dung, animals, and men—and a bringer of light and warmth and comfort. This paradox is at the root of all worship, for we pray to the gods whose wrath we fear the most. Fire, like all other forms of energy, is amoral. Like the Watcher, below, who cares not what it watches, Fire cares not what it burns.

Sacrifices to the gods are routinely burned; this ensures

that the sacrifice will not be reused for any other purpose, of course, but it seemed to the faithful that the rising smoke lifted up to heaven and carried the message of the sacrifice directly to the stars. At the Temple of Solomon in Jerusalem, the Altar of Burnt Offerings was a large platform upon which the fires burned mightily to that hungry God, Jehova. Among the Phoenicians, the sacrifices to Moloch were often children. Jehova was pleased with Abel's sacrifice of livestock and turned up His nose at the vegetable offerings of Cain. Cain, taking his cue from Jehova, thereupon killed Abel. "So, you like blood offerings, eh?" Jehova didn't think that was so funny. Jehova has no sense of humor.

The invocation of the God of Fire—GIBIL or GIRRA—is a prerequisite for the Walking. Since fire is present during the ceremony, it must be treated as any other element of the ritual and properly acknowledged. The Fire God is called a "son of Anu" and is therefore one of the oldest deities in the pantheon. There may be an analogy to the *Ur-Hekau* of the Egyptians: for the Bennu Bird brought *Heka* to the world from the Isle of Fire in the Egyptian myth.

The Watcher

The instructions concerning the Watcher are much more specific, and seem to hint at the remains of an older ritual, some more primitive form of worship to a warrior-deity. The Watcher is described as a member of a separate race of beings, neither Ancient One nor Elder God. These amoral (or, at least, apolitical) entities are supremely powerful, but hire out their services for the price of a monthly sacrifice. As such, they may be considered the mercenaries of the astral plane.

New bread, pine resin, and the grass *olieribos* (nettle) must be burned in a new bowl that has the three signs—the ARRA, AGGA, and BANDAR—inscribed thereupon. Nettle is a noxious plant, and its fumes may cause irritation if inhaled. This plant is common in use among occultists as a protective measure and also in rituals designed to attack or to cause pain, for obvious reasons.

Pine resin—or sap—may be collected fresh from the trees, although it is entirely possible that pine is not intended here but specifically cedar resin from which incenses frequently were made in the Middle East.

As for new bread, that means just what it says. Bread is an offering of grain. These three elements—bread, resin, and nettle—are all plant products and may hint at an early agricultural mythos behind the idea of the Watcher. Note that no blood offering of any kind is required. The burnt offering of the Three Elements is satisfactory and sufficient for the purpose. A blood offering to the Watcher may be made, but that serves to render the Watcher extremely violent and uncontrollable and may lead to the destruction of the magician in a blind lust for blood. This is similar to the reaction one might expect if a fresh kill is given to a dog who has heretofore only consumed packaged dog food.

The Watcher is first evoked by means of a double circle of flour and a new bowl on which are inscribed, as we have mentioned, the three carven signs. There is no altar at this point. The Fire God is conjured in the new bowl after a fire has been lighted therein, and the bowl placed between the two concentric circles, in the Northeast direction. This is to take place at the "darkest hour of the night." There are no other lights or lamps in the vicinity, only the flaming bowl. The robes necessary for this preliminary evocation are simple black vestments and a black cap or headdress of some sort.

There is also the Sword.

The Sword is the symbol of the Watcher, for it represents this force in one of its four forms, the Spirit with the Flaming Sword. The others are: a Dog (like Anubis, a Guardian); a man in a long robe with eyes that never lose their stare; and sometimes as the Enemy, when the magician has failed to make the proper sacrifices or has in some other way offended this powerful spirit.

The Lord of the Watchers is a being that dwells in the realm of the Igigi; in other words, beyond the planetary zones and in the realm of the fixed stars, outside our solar system. The Igigi are also sometimes used to refer to the Underworld beings, the Annunaki, and are numbered as Seven. Again, of course, this is a reference to the seven stars of some constellation, very probably the Dipper.

There is no reasoning with any of the beings called by the generic term "Watcher." They abide by the letter of the Covenant, which in this case is that sacrifices must be made to them at regular intervals or they will not serve; if an attempt is made to evoke them when a sacrifice has not been made within the space of a month, they will turn on the magician instead. That's all we know of the Watcher. There is no higher court to which we can appeal our case in the event that we make a mistake; retribution is swift and final.

The Sword is touched when the Watcher is required for the ceremony of the Walking, and that summons this being to the periphery of the Gate to guard the body of the magician, just as in the Chinese practice described in Part One of this book. It is touched again when it is told to depart. If the Sword is touched at any time during the ceremony, it will depart and cannot be called back during the Rite, so care must be taken not to touch the Sword in any way until a conscious decision has been made to do so.

The one who summons the Watcher must be the same person who sacrifices to the Watcher. Otherwise, the summoner will have opened the Gate between the Watcher's world and the world of the magicians present and all will perish.

To clarify: the destruction of the magician(s) by the Watcher can occur in different ways, although all are swift and irreversible. It can be immediate, during the Rite itself, or can take place as a slow feeding on the body of the magician by the Watcher as all life is drained from the magician. Various illnesses arise, of uncertain provenance, which cannot be cured.

These, then, are the preliminary rituals.

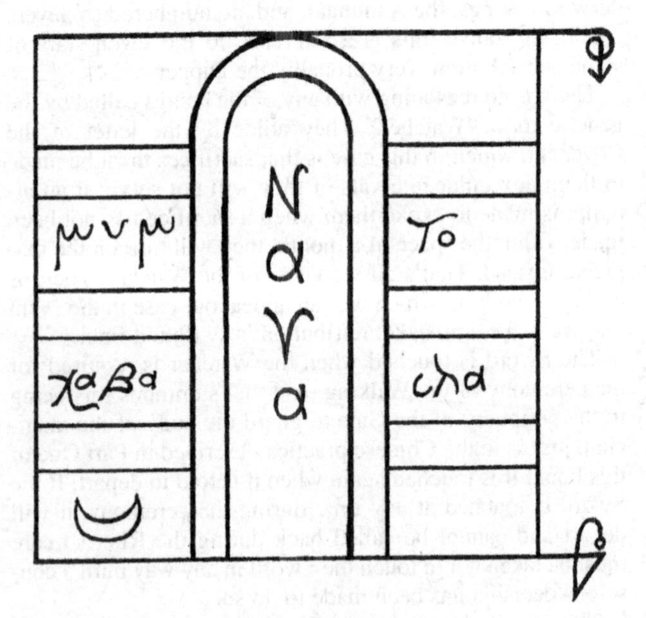

The First Gate—
The Gate of Nanna, called Sin

CHAPTER 8: The First Gate

Once the First Gate has been passed, the Ritual begins to make much more sense than is obvious from the sparse description in the *Necronomicon*. Perhaps the most complex instructions are those concerning the timing of the rituals, and these we have already calculated for you for the next ten years in the Tables of the Bear.

The items necessary for passing this, or any Gate, are as follows:

1. The Sword of the Watcher.

2. The Bowl of the Watcher with appropriate sacrificial items (if sacrifice has not been made within a month).

3. Four lamps for the four quarters.

4. The Seal of the Star to be invoked, in this case of the Moon.

5. Incense.

6. The entire ritual, in your own handwriting, copied faithfully from the *Necronomicon*.

7. An altar, in the North.

8. Representations of the God and Goddess, and sacrificial items. (What these are will be determined by which

deities you select. Reference to any book on mythology should help in case you are at a loss to figure out what your particular deities would require by way of sacrifice.)

9. The brazier wherein the Fire God will be invoked.

10. An offering bowl for the sacrifices.

11. Clothing appropriate to the Ritual.

12. The copper dagger.

13. The Gate itself, drawn on the earth with lime, or barley, or white flour.

The Gate

The Lunar Gate—the Gate of NANNA called SIN—is markedly different from the other six that follow. For one thing, it appears as an actual doorway and is rather symmetrical in design. Similar designs can be found on talismans from the Middle East, and the curious reader is directed to the various works by that indefatigable Orientalist, E. A. Budge, for examples, particularly *Amulets and Talismans*.

In the Doorway, in the center of the Gate, is inscribed the name NANA in Greek letters. The other "words" are indecipherable, but appear to read (from left to right and from top to bottom): MIM to XABA CLA. It is to be noted that there is no letter C in Greek, so its appearance here is something of a mystery, unless an Upsilon or an Iota was intended. There is a lunar crescent in the lower left-hand corner, and the upper and lower frames of the Gate end in a downturning Arrow and the AGGA sign, respectively.

This Gate seems to be modeled after an actual, physical Gate that must have been the entrance to a specific Temple. The horizontal crossbars that run the width of the Gate may suggest the tiers of a ziggurat, but this is conjecture.

The word SIN is of interest here, if only because so much has been made of it in the fundamentalist Christian press.

The word does *not* mean "sin" in the way it is generally understood, but is the *name* of a famous Mesopotamian god of the Moon. Its pronounciation would be approximately "SEEN." The attempt by Bible Belters to use this word as proof that the *Necronomicon* urges one on to Sin should be greeted with the hilarity appropriate for welcoming freshmen to university: the lack of academic background and general common sense peculiar to most freshmen has made them universal objects of ridicule, if only temporarily. One day they will be seniors, perhaps; but for now the Bible Belters get a failing grade in Biblical Mythology 101.

It is appropriate to begin with a "Lunar" Gate, as the traditional arrangement of the planets and luminaries—as common to ancient Chaldea as it is to modern Qabalists—is to begin with the Moon and proceed to Mercury, Venus, the Sun, Mars, Jupiter, and Saturn, in order. "Lunar" experiences include dreams, hysteria, and nervous disorders, as well as growth, periodicity, and other more positive expressions. For the occultist in particular, the Moon is the symbol of the Astral Plane, as it is the first off-world step for us on Earth. It is the initiatory plane, and it is here, at the Lunar Gate, that the vast majority of occultists lose their way, forever.

The temptations of this plane are many. For most people, it is the repository of every inspirational, delusional, ghostly, spiritual, hallucinogenic event that has ever transpired in their lives. The temptation of this plane is to become one of those vague, ethereal types one finds spouting psychobabble on morning talk shows. Many channelers are victims of staying too long on the Lunar level; astral puppets who never progress beyond sitting on the ventriloquist's lap and waiting for the moment the levers are pulled. The mental wards and twelve-step programs are littered with their souls, as the hospitals are with their bodies. Instead of mastering this plane, it has become *their* master; every breeze that brushes across their faces becomes a caress from beyond, every news item a direct message from an entity on

Alpha Centauri. Avoid them like the very plague, except for the brief time it takes to acquaint oneself with the signs of terminal Lunacy so that you may not fall victim to its lures, for they are souls for whom the World has lost its meaning, its joy, and they wait only for death to release them from the icy grip of Life.

The mastery of the Lunar sphere is no small accomplishment, for it will nearly guarantee success in later stages of the Walking. It is worthwhile to note here the peculiar gender of this Gate, as it is properly the domain of a *male* deity, Nanna. This is not unusual among the Indo-European cults, and elements of this gender classification still exist in some of these tongues, such as German. In terms of the *Necronomicon*, the Moon is Male: a kind of psychopomp who will lead you into the darkness like a careful shaman. Do not linger too long in his company for, as it declares in the Invocation of the Nanna Gate, his "thought is beyond the comprehension of gods and men . . ."

When this Gate is open, it is the equivalent of opening the Ur-Gate between This World and the Other; the veil that separates the two parts for a moment when this Gate is opened, and anything becomes possible. In a sense, this is the most important Gate of all. If you are successful here, you will have established a foundation in the Otherworld from which to build your own astral edifice.

The Seal of the Gate

This is composed of three horizontal bars (note the similarity to the three horizontal planes on the diagram of the Gate itself). The topmost bar shows the familiar glyph of the eight-pointed star that denotes "deity" in Sumerian cuneiform. As mentioned before, this probably represents the Seven Stars of the Dipper plus the Pole Star (which would not have been Polaris at the time of the ancient Sumerians, but would have been at the time of the Mad Arab), or the Eight Stars of the original Dipper, that is, including Arcturus.

On the central bar is the familiar crescent Moon, turned in such a way as to imply horns.

The third, and lower, bar is composed of three glyphic elements that it is impossible to decipher and which may represent certain secret formulae of the Gate itself. The Seal should be engraved on silver on the thirteenth day of the Moon. No ray of sunlight should ever be permitted to strike the Seal, lest it lose all its power in that moment. It should be put away, in darkness, wrapped in silk.

Other Information

The number of this Gate is, of course, Thirty. This means that the magician will walk—circumambulate—the Gate a total of thirty times, beginning in the North facing the altar, and proceeding clockwise until all thirty rounds have been made. This is the Mad Arab's "Pace," and it is important to take this part of the ritual very seriously and not to rush through it.

The incense should be camphor.

It is possible to visualize the Lord of this Gate as he is described in the *Necronomicon*, that is, as an elderly but powerful man with a long beard and a rod of lapis lazuli. He is called the Father of the Gods and the Eldest of the Zonei. The rod of lapis might have represented a measuring stick of ancient times, as the words "moon" and "measure" have the same root in many languages. The colors that seem to represent this Gate are silver and deepest, darkest purple.

It is possible that the Lord himself will give you the Word you need, but do not call upon him unless you are experiencing nothing within the Gate at the time of the ceremony. If, after visualizing this Lord, you still receive no Word, simply proceed to the rest of the ritual and close down the Gate as per the instructions. You will receive the word within a few days.

The symbols of the four quarters are given in the pages of the *Necronomicon*, and you may feel it necessary to have these inscribed on the same ground in their proper orientation before the start of the operation. It is possible, and generally more convenient, to have these Quarter—or Earthly—Gates drawn or embroidered on large sheets of parchment or silk, respectively, for they can therefore be reused many times and acquire something of a charge over time. Others, however, feel that this is not as powerful as redrawing them each time, much the same way Native American shamans in the southwestern part of the United States draw complex occult symbols on the ground in many colors for specific purposes, or like the *vévés* of the Haitian *voudoun* ceremonies. Whether you inscribe these symbols on the ground or on more permanent surfaces is largely determined by whether you are a European-style ceremonial magician or a New World–style shaman.

The Procedure

With all of the above in mind and at hand, we can go over a typical opening of the First Gate step by step, with some additional suggestions based on the Daoist analogue.

STEP ONE: THE SEAL

The Seal of the Moon is to be inscribed or engraved on silver on the thirteenth day of the Moon. This day may generally turn out to be the actual Full Moon or a day or so before the actual Full Moon, depending on which time of day the Moon becomes, technically, Full.

The making of this Seal could conceivably take place on the same day as the Walking of the Lunar Gate, but the magician may find that there is too much with which to be concerned on the day of the Walking. Rather, it is advised to make the Seal on the previous Full Moon night (i.e., a month earlier) and keep it wrapped up in silk the entire time before the day of the Walking. As Crowley has pointed out, the time such an object spends wrapped and put away is the most crucial time in its consecration. Psychologically speaking, during this time the object is "forgotten"—i.e., sinks into the unconscious mind, where it accumulates the necessary psychic force—only to be "remembered" for the ceremony when it is brought up, out of its unconscious location, and attached to various psychic mechanisms (identification with personal "lunar" associations unknown to the conscious self), which will prove useful in the Rites to follow.

STEP TWO: THE PURIFICATION

According to the rubric, the magician must have kept the period of one Moon "pure"; we may calculate that from the date the Seal was made, in our example, to the date the Walking is attempted. Sexual intercourse is forbidden during this time, unless the magician knows and is able to perform the act known as *karezza* or some form of *coitus interruptus* in which orgasm is not achieved. Note that it is permissible for the partner to achieve orgasm in this case, except where the danger of sexual vampirism exists. This is not a subject for this book, and the above hints should suffice for those who know what is actually being discussed.

Each day of the Moon, the magician must call upon the God at dawn and the Goddess at dusk. It is also useful to summon the Watcher during this period, to accustom oneself to the ritual and to perform the necessary sacrifice so the Watcher is satisfied on the day of the Walking itself.

All the clothing, incense, and ritual paraphernalia should be collected during this time if not before. The Ritual should be written out in the magician's own hand during this time, including the Invocations of the Fire God, the Watcher, the Four Earthly Gates or Watchtowers, and the Invocation of the Gate, as well as the part of the *Magan Text* that treats of Inanna's Descent into the Underworld, and detailed instructions as to the order of the Ritual. Although it is highly recommended that all Invocations be memorized completely, it is imperative to have a copy of the ritual with you in the Gate, as it has been known for memory to fail at the crucial moment.

STEP THREE: THE FAST

For the seven days up to and including the day of the Walking, the magician must not eat meat of any kind. This is the beginning of the fast. Four days later, the fast extends to include all solid food whatsoever, allowing only water to be drunk from this point on. During these final three days, the magician is to call upon the Three-in-One of Sumerian ritual: ANU, ENLIL, and ENKI. All of this, of course, is in addition to the daily invocations of the God and Goddess.

STEP FOUR: THE CEREMONY

At this point, the day of the operation itself has arrived, being the thirteenth day of the Moon. That day, preferably after the twilight Invocation of the Goddess, begin to inscribe the sacred drawings of the Lunar Gate and the four Earthly Gates (if desired), and place the four lamps in their proper locations.

Set up the altar in the North, with the two Deities.

Set up the brazier before the altar, and have the incense ready at hand.

When all is in readiness, and at the hour of midnight, approach the Gate "with awe and respect."

Light the fire in the brazier with the Invocation of the Fire God.

Throw incense on the fire, and make the appropriate sacrifices to the twin Deities.

Then light the four lamps from that primary fire, invoking the four Gates of the Earth.

Then recite the Invocation of the Watcher, raising the copper dagger of Inanna and "thrusting the Sword into the Earth at its station."

Take the Seal of the Gate that was made a month previous, unwrap it in the presence of the twin Deities, and whisper its name softly upon it.

Recite the Incantation of the Walking, "loudly, and in a clear voice," as you walk around the Gate clockwise the prescribed number of times; in this case, thirty times. This is the core of the ritual. It is not necessary to recite the Incantation thirty times, only as often as it is felt to be necessary. Once is quite enough, but the final phrases in Sumerian seem to bear repeating, as they tend to throw the conscious mind of the magician into a strange fugue state in which the Lord of the Gate can more easily make his Word known.

Once the thirty circumambulations have been accomplished, the magician now falls down before the altar—taking care to look in no direction except up above the altar—and waits to hear, or possibly see, the Word of the Gate. The Gate may be seen to actually open in the air before the magician at this time, and various other sights may take place. The magician should take conscious note of everything that is seen, heard, or otherwise experienced *within the Gate*. Nothing that takes place in any other part of the Temple area should be recorded.

Once the vision has passed, the magician will experience a "falling" sensation and will find that the Temple has been, in a sense, reentered. All other sensations will have stopped. Everything will be peaceful.

Thank the twin Deities of the altar.

Touch the Sword of the Watcher so it might depart.

Recite that part of the *Magan Text* that deals with Inanna's Descent into the Underworld. This serves to remind the magician of the seriousness of the rituals that are being performed as well as to act as a kind of protective incantation for the body and soul of the magician once the Temple area is abandoned.

Extinguish the fire. The ceremony is over.

Retire at once to a safe place and write down everything you experienced during the ceremony, taking care not to commit to paper anything that might have taken place around the edges of the Gate, for those are the actions of the dread *idimmu*, demonlike beings that are attracted to occult operations and try to feed off the power that is expended. To write of them and their machinations is to preserve their *actions*; this opens a Gate for them into this world that is difficult to close.

Daoist Analogues

This ceremony is equivalent to approaching the first Star of the Dipper in the Chinese system: the one called "Clarity of Yang." This is the star at the lip of the bowl. Performing this ritual at midnight—the hour of "rising Yang"—when the full Moon is at its height, unites the "rising Yang" energy of the magician with the reflected solar light from the Moon. It is an alchemical event, this mingling of the two "breaths," of Yang and Yin essences, and the Moon is a convenient symbol for this first Star. Of all the nighttime objects in the sky, it is, of course, the brightest and largest, and therefore composed of the greater "Yang" qualities in terms of size and light. In terms of Daoist imagery, one uses the rays of the Moon to travel to the stars of the Dipper in much the same way one might use an escalator.

Lapis lazuli—which forms the rod of the Lord of the Nanna Gate—is also a substance we find in the palaces of the Stars of the Northern Dipper in Shangqing Daoism: lapis lazuli and a "watery substance" form the structure of these palaces, which relate them very closely to the Moon. It would seem from this information that the Moon sketches the outline of the experience we will have in this, and succeeding, Gates. The colors silver and purple; the lapis lazuli rod; the watery essences found in the "palaces" of the Seven Stars; all these are Lunar correspondences that are found both in the *Necronomicon* and in the sacred texts of the Daoist alchemists. The Moon, therefore, is a symbol of a class of experience that is echoed throughout the remaining Stars. In this way, it is truly the "most ancient" of the Gates, and entering it the most important step the magician takes, for by doing so there is no return.

The Three-in-One are an essential part of the Daoist ritual, and they represent the interplay of the essences that are the basis of Chinese—and Western—alchemy, being Mercury, Sulfur, and Salt. In the Pace of Yü there are elaborate mechanisms for stimulating the growth and purity of these substances and for achieving their harmony and balance in the astral, as well as the physical, body. We will not go into this detail now, as it deserves a book-length treatment of its own in light of Western alchemy and ceremonial magick. The circumambulations as practiced by the Daoists are described in various ways by different authors. One method that seems most appropriate for our purposes is the repeated clockwise circumambulation that takes the form of an inward turning spiral. Some of those who have already performed the Walking report that this spiraling towards the center of the Gate seems to work quite well. Indeed, in some cases it seems to occur naturally in the course of the ceremony without deliberate intention.

The summoning of spirits of the Four Quarters is part of the Chinese ceremonies of Walking, as well. To assist the Western magician in establishing these "four pillars," an imitation of the Daoist method may prove useful. These four quarters may be called "down," as if they were stars in their own right, or summoned "forth," as if they were forces within one's own *body*. Either of these visualizations works quite well, although the sensations differ slightly.

After these four invocations have been performed, it is useful at that point to visualize the stars themselves descending upon the magician, clothing that individual in the stellar radiance. It may help to imagine oneself as a center of gravity, drawing the astral stuff towards one from every corner of the universe. This concept of the Self as a center of gravity has already been discussed by Tufts University professor of philosophy Daniel C. Dennett in his book *Consciousness Explained*. While this author does not agree with every point of view Dennett expresses, there is much food for thought for the serious occultist within the pages of that book.

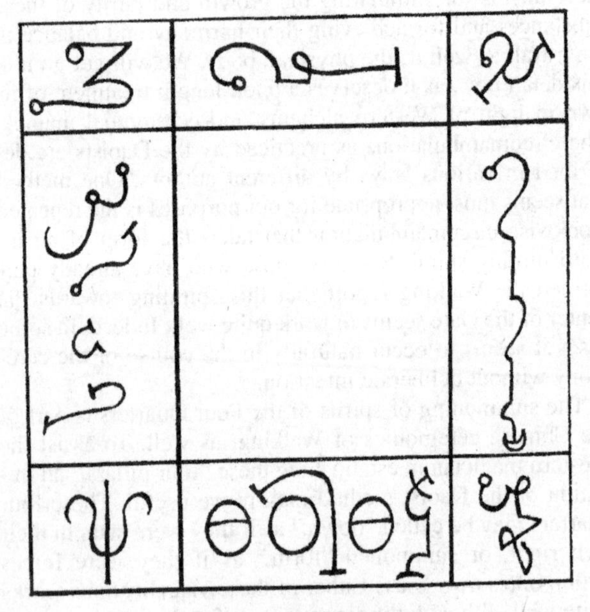

The Second Gate—
The Gate of Nebo

CHAPTER 9: The Second Gate

Allowing at least the space of one month to pass before starting the Second Gate, the Ritual is identical in all respects to the First save for the correspondences particular to this Gate, which is the Mercury, or NEBO, Gate. Therefore, we will restrict ourselves in this and the following chapters to a discussion of the differences and not bother to repeat the entire ceremony each time.

The Gate

This Gate is designed more conventionally than the NANNA Gate. It is a simple Square with a border on all sides, like a frame. Within the boxes formed by this frame are various sigils and glyphs, none of which are in a known language or have identifiable analogues, save for the figure in the lower left-hand corner, which is either the Greek letter Psi or a stylized trident or caduceus. The latter would be conventionally appropriate for the associations given to the planet Mercury in its aspect as Hermes, but that is not necessarily the case here.

The bottom-central bar contains three circles, two of which are bisected by a straight horizontal line, and another line that seems to arc around the first two. Possibly, this refers to some peculiarity of Mercury's orbit or period.

As always when dealing with occult symbols on talismans, amulets, magic circles, etc., when the meaning or representation of a symbol is not known or understood, the symbol must be copied exactly as shown. The errors that can obtain through sloppiness or carelessness can cause grief in the future. Many of the original Angelic Tablets of Dr. John Dee and Edward Kelly were miscopied and misprinted, causing no end of troubles for the Golden Dawn; many of the Hebrew grimoires are also full of copyist's errors. For this reason one finds the strict admonition in the *Necronomicon* and virtually every other grimoire: copy this exactly as you see it, changing nothing, no matter how small or slight a change it appears. Those who have studied Hebrew are aware of the slight difference that exists between, say, the letter Hé and the letter Cheth, or between Daleth and Resh: differences that, to a Gentile eye, seem insignificant but that change the pronounciation, gematria, and meaning of the word entirely. One thinks of our letters U and V, or I and L. The same is true of magickal glyphs. Mistakes are *not* okay.

The Seal of the Gate

Contrary to the Gate itself, the Seal is rather asymmetrical and strange. This penchant for off-center lines and angles is a peculiarity of the *Necronomicon* among grimoires and is probably a key to its singular power. The intention of the Mad Arab seems to be a demonstration that the "planets" are not what they appear to be: that the Moon is not *really* the Moon, and Mercury not *really* the planet Mercury, but that a space is indicated at *odd angles* to our normal perception of these bodies. The sigils of the *Necronomicon* force our eyes away from the center and to the sides, keeping us off-balance, forcing our consciousness to compensate for the lack of psychic equilibrium. A comparison of the *Necronomicon* seals to those of any popular medieval grimoire will show how startling is the difference in outlook and, presumably, in intention. The *Necronomicon* intends to take us somewhere the other systems do not.

Other Information

The Lord of this Gate is NEBO. He is also quite old, and bearded. He has the traditional function of Mercury in Western mystery cults, and that is as the God of Wisdom and Science. In this case, however, he wears a crown of one hundred horns. This may refer to some observable quirk of the planet Mercury, perhaps, such as its orbit or the sum of its conjunctions with another planet within a specific period of time.

The color is blue, and the metal is Quicksilver. The Seal must be inscribed on perfect parchment or on the broad leaf of a palm tree.

The number is Twelve, which is the number of circumambulations required for the Walking itself.

Daoist Analogue

This is the Second Star of the Dipper in the Pace of Yü. Its name is "Essence of Yin," and is the outer bottom edge of the Dipper's bowl.

It forms, of course, a polarity with the first Star, the "Clarity of Yang." Once this Gate has been passed, and the Word received, a subtle mechanism has been established between the twin psychosexual poles of the magician's being. That this is what is intended is clarified by the name of the Third Star in the Daoist context: the True (or Real) Person.

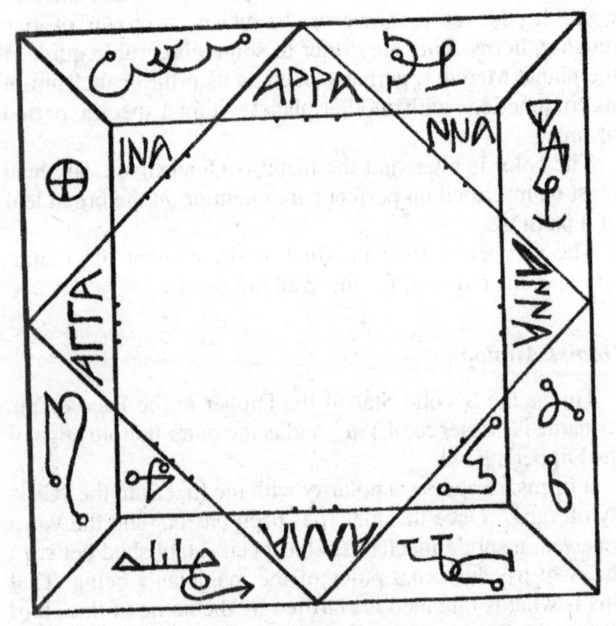

The Third Gate—
The Gate of Inanna, called Ishtar

CHAPTER 10: The Third Gate

This is the Venus Gate, the Gate of INANNA called ISHTAR and, later, ASTAROTH and ASTARTE. We remember that it was INANNA who descended into the Underworld, thus providing a model and a precedent for our Quest. Although the association—in Sumerian mythology and in the *Necronomicon*—is with the planet Venus, if we consider Her the Third Star of the Dipper, we find that She occupies the place called the "True One" or the "True Person." This Star is located directly before the Star called the "Underworld" or the "Profound Darkness." Thus, we have once again realized a great deal of agreement between the two systems.

The Gate

This diagram approximates very closely the mandalas of India in its repeated symmetries of squares at right angles to each other. In the diamondlike central square we find the words ARRA, ANNA, ALLA, AGGA written in Greek letters in the triangles formed by the intersection of the squares. ARRA and AGGA, of course, are the names of the Pentagram and the Elder God signs, respectively. ALLA could be a crude approximation of the name of God in Islam. As for ANNA, that name could be subject to a volume of exegesis itself, for it combines the name of the first God, AN, with its negative, NA. AN is a word we find all over the world in various contexts, and was not unknown among the ancient Egyptians and Greeks as a name of God. Indeed, INANNA and NANNA may both be variants of this name, and the *Necronomicon* itself clearly states that NANNA and INANNA partake of a "shared essence." This essence, of course, must be AN and must thereby refer to ANU of the Sumerian Three-in-One. This is a very provocative and suggestive statement, of course, and the above references should serve anyone serious enough to pursue the proposed correspondences further. Interested readers are referred to the Bibliography for more clues.

The word INANNA also appears, in the two corner triangles, and the AGGA itself appears in the lower left-hand corner. The symbol in the lower right is of unknown provenance.

As for the other glyphs and sigils, they are equally obscure. The circle with the cross within it may be a symbol for "city," but this is only conjecture.

However, one should draw attention to a curious fact about this Gate, and that is the presence of three small lines on each side of the inner octagon formed by the intersecting squares. These are similar to the lines one finds on the Sumerian hieroglyphic for "Temple," BAR, and may have once indicated steps or a passageway or even separate Gates. In the case of the Seal of the INANNA Gate, there are then a total of twenty-four of these lines. The figure eight multiplied by three is one we have already discussed in an earlier chapter and refers to the eight "effulgences" multiplied by the three "cinnabar fields," to give a total of twenty-four "perfected immortals." Numbers were sacred to the ancient Sumerians, as they were to virtually every cult since then. The precise number of twenty-four lines would have indicated something and not been the result of random artistic expression or chance. Whether the number twenty-four had the same significance for the Sumerians, or the Mad Arab, as it had for the Chinese Daoists, is anyone's guess at this point due to the paucity of written records on this matter from the former.

As can be seen at once, we have progressed from the relatively obvious "doorway" of the First Gate, to the more stylized representation of the Second Gate, to a full-formed mandala for the Third Gate, thus expressing a progression of experience as well as of architecture. We have attained a degree of perfect symmetry in all directions by passing the INANNA Gate. We are armed with Her armor, and are therefore "similar to the Goddess."

The Seal of the Gate

The concept of diagonals that was introduced in the Gate diagram itself is echoed in the Seal of INANNA with a pair of crossed diagonals in a square, forming four triangles.

In the right-hand triangle we find the eight-pointed Star referred to many times as the symbol of divinity, and as a reference to the Stars of the Dipper (plus either Arcturus or the Pole Star).

The lower triangle may represent a planetary or stellar orbit, but as with all the other mysterious sigils, who knows? The odd shape in the left-hand triangle is suggestive to this author of the Egyptian adze as it was written in some later hieroglyphic samples, and, if so, would reinforce the idea that the eight-pointed Star in the opposite triangle is, indeed, the Dipper for the adze is the Dipper-instrument and Dipper-shape.

The figure that joins the left triangle to the upper triangle is unknown, as is the purpose behind the horizontal crossbar in the upper triangle.

Other Information

The number of this Goddess is, of course, Fifteen. In a way, that is similar to the number of the *Necronomicon*, which, adding up its letters in Greek, equals 555, a number of "obscurity" according to *Liber 777*.

The Goddess is described as a Goddess of Passion, both of Love and of War (an obvious reference to the planet Venus as Morning Star and Evening Star). She appears as a beautiful woman in the company of lions, and sometimes appears in armor when she is known in her aspect as the sister of Ereshkigal, Queen of the Underworld, and her conqueror.

Her color is white, her metal Copper (on which her seal must be engraved when the planet Venus is in exaltation).

It is said in the *Necronomicon* that when NANNA and INANNA are in agreement—their planets being in auspicious aspect to each other—it is "as two offering-cups spilt freely in the heavens, to rain the sweet wine of the Gods upon the earth." We are reminded of the Tarot Trump, *The Star,* of which Tzaddi is not.

Daoist Analogues

Many of this Gate's Chinese associations have already been mentioned, but we should examine one or two more.

The *Necronomicon* has a bizarre phrase which is worthwhile to quote in its entirety:

> BUT KNOW THAT INANNA TAKES HER OWN FOR HER OWN, AND THAT ONCE CHOSEN BY HER NO MAN MAY TAKE ANOTHER BRIDE.

One cannot help wondering if this is not analogous to the "Celestial Maiden" familiar to the Chinese alchemist from the solitary practice of *ou-jing* or "paired effulgences," in which union with the Celestial Maiden is attained through meditation and Daoist yoga techniques. The celibacy of these alchemists was never endured in ignorance or avoidance of the sexual aspect of human life, but sexuality was actively incorporated into the ascetic vocation through a variety of methods. This is in sharp contradistinction to the Roman Catholic state of celibacy for priests, which has no such evolved methodology for actively using sexual energy towards mystical ends, but instead is a denial or resistance of sexuality as an "evil impulse."

The Chinese characters for this Star are *zhen-ren*, or "True One," which may be an allusion to the similar sounding and almost identically written *jen-ren*, or "perfected immortal." If so, then our discussion of the twenty-four bars on the diagram of the Gate as representative of the twenty-four "perfected immortals" suddenly becomes quite startlingly valid.

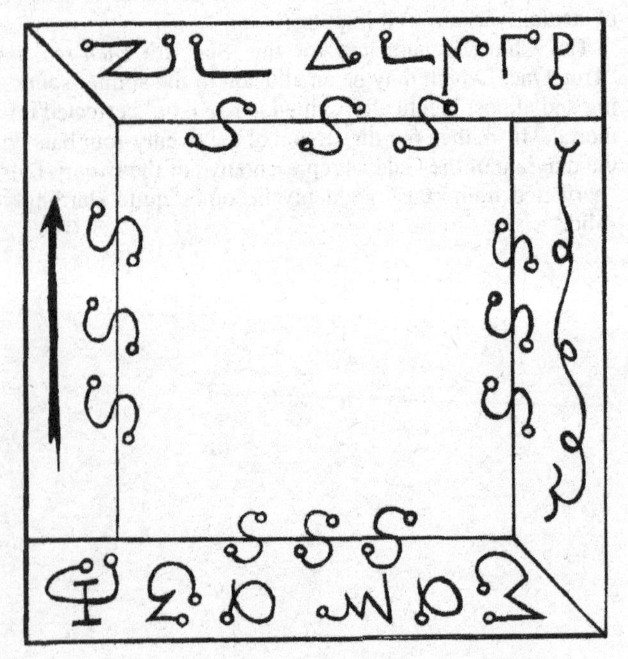

The Fourth Gate—
The Gate of Shammash, called Uddu

CHAPTER 11: The Fourth Gate

This is the Gate of the Sun, SHAMMASH, also called UDDU, Shammash being the more Semitic or Babylonian nomenclature and Uddu being more properly Sumerian. In this cosmology, the Sun is seen as the son of the Moon. Very little is said concerning this Gate and its Lord, save for the usual references to Gold and Light and Flames.

The Gate

The design of this Gate is unusual in that the "picture frame" style has been altered in an Escherlike fashion so that it is impossible to obtain a clear sense of perspective. There are the usual indecipherable glyphs and sigils, along with an upward-pointing arrow in the left-hand bar.

The four sets of S-triplets along the inner walls give us a total of twelve S's, which may be a reference to the twelve signs of the Zodiac through which the Sun moves in a year, divided into four quarters for the four "platonic" elements: fire, earth, air, and water.

The Seal of the Gate

The Seal is, in a way, more interesting than the Gate. It shows a simple, conventional Gate composed of two uprights and a lintel, rather like the entrance to a mine. Standing, or floating, in the center of the Gateway is a being of eight arms, which is probably representative of the Sun and its rays. The interesting thing about this glyph, however, is the number of rays—eight, as in the symbol of Divinity and its cognate, the Stars of the Dipper—and the wavy nature of these rays rather than the straight or jagged solar rays that are familiar from Middle Eastern ikonography. The overall effect is of a tentacled Being, perhaps malevolent, standing at the threshold of the UDDU Gate.

The three signs above the Gate are also of unknown provenance and meaning.

Other Information

The number of this Gate is Twenty, for twenty circumambulations. It should be pointed out that these numbers are of Sumerian origin and do not seem to refer to any specific quality of the planets to which they refer, but are rather based on a peculiar numerology of the Sumerians in which Sixty is the perfect number and all other planetary numbers are equal divisions of Sixty. In this sense, the Sun is one-third of the Perfect Number, whereas the Moon is one-half, and Venus is one-quarter. The one exception is Mars, whose number is Eight. The reason for this anomaly is obscure. Sixty divided by Eight gives us Seven and a half, a number with no discernible relevance to anything.

The Seal should be inscribed on Gold, and this is probably one reason why, as the Mad Arab says, many individuals stop at the Inanna Gate and go no further! Actually, the Seal is relatively simple to inscribe on a small square of that metal, so the cost of obtaining sufficient Gold to enable one to attempt this Gate is not prohibitive. What *is* prohibitive is the very nature of the Gate itself, for a paradox lies buried within this particular mine shaft, as we shall see.

Daoist Analogues

According to the Chinese, this Star of the Dipper has the somewhat sinister connotation *xuan ming*, or "Underworld," "Black Obscurity," "Mysterious Darkness," etc. This is indeed a strange appelation for a Gate that, in Sumeria, was ruled by the Sun.

Remember, however, that this ritual is performed at night, at midnight during a Full Moon when the Sun is at the opposite end of the Earth from the Moon. It inhabits the North, not the North of the Pole Star, obviously, but the North of the Underworld, the "mirror image" of the Seven Celestial Gates. In astrology we know that the North of any chart is the Nadir, the lowest point. It is where the Sun goes on its nocturnal journey below the Earth in Egyptian mythology. The North, therefore, is the Land of the Midnight Sun in truth; it is the place of rebirth also, for when the Sun leaves "midnight," it begins to grow and gather around itself the powers and abilities that it shucked off on its downward journey into the Netherworld until it is reborn at the moment of sunrise.

Hence, the Daoist associations once again ring startingly true, for although the author knows of no document specifically relating each of the seven "planets" to each of these star names in the Dipper, we have before us an initiated interpretation of the Sun and the Underworld that is consistent with Egyptian and Sumerian mythology.

The Fifth Gate—
The Gate of Nergal

CHAPTER 12: The Fifth Gate

This Gate is the domain of the Red Planet, Mars, and of the Sumerian God NERGAL. The martial qualities of this planet were not lost in translation, for the *Necronomicon* calls his seal "a sharp sword."

In the Fourth Gate, we discussed the place of the Underworld as the womb of the Sun. At the Fifth Gate, the Divine Birth takes place in a frenzy and a clamor.

The Gate

We have returned to a very symmetrical framework for the NERGAL Gate, which is probably just as well. One can only wonder in horror at what might transpire should the violent forces of Mars be constrained to flow through an asymmetrical space, thus distorting with primeval power the very framework of the Universe.

According to the Qabalists, just such an event transpired in ages past, which resulted in the creation of the *qlippoth*, the *shells*, when the "vessel" containing the Martial energy at the sphere of *Gevurah* shattered. The shards of *Gevurah* are the shells and demons of occult lore, the dread qlippoth. That is another reason why they fear the Five-Pointed Star, for Five is the number of *Gevurah* in its whole, completed state. One may find many instances in the lore of demonology where the Five-Pointed Star is shown broken, to represent evil forces.

In the symbolism of this Gate we find this very Pentagram, whole, and with attachments. It bears a horned crown, which may simply be the crescent Moon, and two objects suspended from the lower arms of the star, of uncertain meaning. (The author has received communications from high priestesses of Gardnerian-style Witchcraft in which they have signed their names accompanied by a pentagram with a flame at the top point to denote their rank.)

Above the central square can be seen the BANDAR, the sign of the Watcher. When this sign appears, we know we are treading on dangerous ground. The BANDAR represents force in its most naked, most amoral state: blind energy that must be controlled else it destroy the user.

As for the other symbols, they fall into the same category as the vast majority of *Necronomicon* sigils: incapable of definition. The only remaining symbol we can translate with confidence is the one in the lower left-hand corner, which is obviously that of a knife or sword. The symbol in the upper left-hand corner may be a constellation, as well as the symbol in the left-hand inner frame; but these constellations are unknown to modern astronomy and may refer to an alternate arrangement of stars.

The Seal of the Gate

Where our Gate was symmetrical, our Seal is not. Again, we have a deliberate asymmetricality that points to a sinister application of the forces of this Star. The seals in the four quadrants are all sharp edges and points. There can be no doubt as to the warlike nature of this Gate.

The left-hand symbol seems to be a collection of branches or sticks held together to make a torch, a common enough implement in all primitive cultures, and in this place may symbolize the torch of Hecaté, held aloft during the descent into the Underworld.

The upper symbol seems at first glance to be a crown, but

may have alternately represented an instrument of war, such as an axe. The other two symbols are mysterious.

Other Information

NERGAL has the head of a man on the body of a lion, and carries a sword and a flail. Although a sword is a natural enough weapon of war, the flail is usually used to inflict pain without killing; that is, it is an instrument of torture, of punishment, and nothing else. As the *Necronomicon* states, Nergal "was once thought to be an agent of the Ancient Ones, for he dwelt in CUTHA for a time." CUTHA, of course, is the Underworld from whence the name KUTULU or Cthulhu, "Man of the Underworld."

His color is dark red, and his metal is Iron, as is that of Set. His number is Eight, an odd number with reference to the others, which are all, as we have said, equally divisible into Sixty, but perhaps an insistence on behalf of the mythologians that Nergal—for all his unpleasantness—really does belong to the realm of the Eight-Pointed Star and is somehow crucial to the entire Operation and not to be avoided.

Daoist Analogues

Here again we have an excellent correspondence with the Seven Stars of the Dipper, for the name of this Star translates as "Red One." Mars is, of course, the Red Planet; there can be no accident that "Red One" is the name of this star in this exact sequence. As the Shangqing alchemists have stated, the Dipper is really the essence of the seven planets, which are grosser forms of the Stars.

Red is also the color of cinnabar, the sacred element of the alchemists of every country from China to England. During the Pace of Yü—according to one description—at the crucial moment a cloud of red smoke comes down and covers the alchemist, who then ascends to the Northern Dipper suspended by this red cloud.

**The Sixth Gate—
The Gate of Lord Marduk**

CHAPTER 13: The Sixth Gate

This is the Gate of MARDUK, the Jupiterian Deity who is responsible for having defeated Tiamat—Leviathan—in an ancient war that resulted in the creation of the universe as we know it, and of humanity in particular (from the flesh of the slaughtered Monster and the breath of Marduk himself). Therefore, Marduk is our direct connection with divinity, for he, more than any other God, is our Parent: the first in the line of ancestors of the human race.

A particular aspect of this Gate is, once having passed it, one may then call upon the famous Fifty Names of Marduk. These were bestowed upon him by a grateful council of Elder Gods, and are usually treated as fifty separate and distinct entities. They form the structure of a separate book within the *Necronomicon*—the Book of the Fifty Names—which has all the appearance of a standard, medieval grimoire replete with sigils and words of power and divine names and the appropriate abilities of each "entity."

The Gate

The design of this Gate is deceptively simple, for it is nothing more than a frame around a central square. There is a strange diagonal drawn in the upper left-hand corner that appears to connect the inner square with the outer frame.

Just below that diagonal is the AGGA symbol, the sign of

ENKI the Master Magician, Lord of the Abyss above the Underworld and therefore a symbol of the subconscious mind as opposed to the unconscious mind: i.e., the twilight layer between conscious mind and unconscious mind. In this fashion, ENKI—once triggered—acts as a kind of psychic traffic cop, directing signals between both "minds." He can also be thought of as a psychic placenta, a device for receiving nutrients from the outside world and apportioning them to the inner world according to a complex system of needs and desires, of correspondences and the alphabet of symbols.

Trailing around three-quarters of the frame is a vinelike object with tendrils. Opposite the AGGA is the BANDAR once again, and we find we have not lost the Watcher. In fact, directly above the BANDAR is a drawing of an Eye. This may signify consciousness, and in addition it may also be a reference to the Egyptian "Eye," which is a complicated subject beyond the scope of this book but can be described as the searing power of the Gods defined as Light and Vision. One can only suspect what it means in combination with the BANDAR.

The Seal of the Gate

This is a busy instrument, replete with arcane etchings, a few of which are intelligible. The BANDAR once again appears on this Seal, as on the Gate itself. The Eight-Pointed Star appears on the opposite side. The figure in the center appears to be a being holding a sword aloft in one hand and a shield in the other.

Other Information

Where Nergal is blind animal force of battle and Inanna the passion of the righteous cause that motivates even those who would otherwise abhor bloodshed and violence, Marduk represents the victorious general in charge of the troops: he knows the wages of war and the price of victory; he is also the intelligence behind the strategy and tactics that are employed in warfare. For him, the dubious joys of righteous anger and blood lust are strangers; his function is the arrangement of the complex elements of battle—food, armaments, wagons, horses, foot soldiers, spies, fortresses, weather, terrain—so that they may achieve the desired end. In this, he is the Magician of War, and like all magicians, he is also a Creator.

Where Nergal represents Will—"pure Will, unassuaged of purpose"—and Inanna, Desire; Marduk is the Law. This Law is not so much the law of courts and decrees, but the Law of science, the lineaments of the created universe. Through the first five Gates we have become initiated into the use and sense of various Forces; in the Sixth Gate we become masters at manipulating all of them, at mixing them to produce various effects. We have passed through the Gates of Hell; Marduk has mastered them.

His color is Purple, his metal is Tin or Brass. (These are traditional Western associations with Jupiter.) If Marduk seems a bit more benign to you than Nergal, remember that in Marduk, Nergal achieves his apotheosis.

The Number of Marduk is Ten, and thus the number of the circumambulations.

Daoist Analogue

"Northern Bridge" is the name of this Star, and it leads directly to the Celestial Gate itself. This may find its echo in the Seal of Marduk, which shows a central bar occupied by a Warrior, with another bar at right angles leading off of the former. This may indicate a bridge over which we must pass. In the Daoist scriptures there are said to be two "invisible stars" in the Northern Dipper. One of these hangs off this Star as it is shown in the old diagrams. Thus, the "bridge" concept may be appropriate to the Seal as shown in the *Necronomicon*, as well. The Bridge that leads off to the right of this Seal may very well lead to this "invisible star."

The Seventh Gate—
The Gate of Ninib, called Adar

CHAPTER 14: The Seventh Gate

This is the final Gate in our series. It is the domain of the Sumerian NINIB, also known by its Semitic tag, ADAR. This is the outer limit of most initiations in our world, and corresponds to Saturn, the outermost planet known to the ancients (we think). To have passed this Gate is to have achieved a wonderful thing, for it will have demanded patience, persistence, and moral courage. Consistent behavior is a cherished commodity among humans when it doesn't degenerate into addiction or other forms of living death. Generally, there is some equally consistent gratification in store for those who are constant: drug-induced euphoria or multiple orgasms, for instance.

But in the practice of ceremonial magick, consistency often seems to be its own reward. The gratifications enjoyed by most people do not obtain in the pursuit of spiritual goals. For most of us, spiritual gratification arrives a day late and a dollar short. To the uninitiated, it is an end, not a means to an end. To the initiated, the means and the end are one in the same. The map is the destination.

This is difficult to experience for newcomers to all of this. There is always an initial *frisson* of something when one first steps into the world of occultism. Usually that involves an eerily correct prediction by a psychic or the vision of a ghost or some other unnatural phenomenon. That is the First Gate beckoning.

To go beyond these experiences—to have them abandon you halfway through the journey—is a condition that most "seekers" cannot accept. They discover that their "spiritual quest" was nothing more than a search for more spooky thrills, and they gaze into Crowley's "magick mirror" and see muck. They return to their daily activities, busily convincing themselves and each other that "there's really nothing to all that magick stuff," when their experience has shown them that there is really nothing to *themselves*. These poor souls think they can return to a normal life and leave their one glimpse of their spiritual potential behind them forever.

Fools.

The Gate

We have before us another conventional Doorway, with inscriptions on the sides and overhead. This telescopic effect is obviously intended to communicate a passageway, a feeling of going *deeper* into something, somewhere.

Around the outer doorway we have some Greek words separated by a Pentagram. We have already discussed the Pentagram at some length, so no further clarification is required at this point.

The word running along the left-hand side seems to be XAMODAIMOM, and the word running down the right-hand side seems to read MAXOSIGIGI IA. The last word has a crescent Moon symbol separating the suffix IA from the rest of the word. This seems to be a symbol rather than a letter, as the ends of the crescent end in circles, which none of the other letters do. This would appear to rule out a stylistic affectation and instead mean something entirely different: a pair of horns, obviously, or the Moon itself.

XAMODAIMOM has no Greek equivalent. For a while this author thought that the Greek word for evil spirit, KAKODAIMON, might have been intended, but he is not certain. MAXOS IGIGI IA might be a halfway intelligible phrase, if MAXOS is an attempt to render the Greek word *mahimos*, "warlike," or possibly *maksi*, "together with," or even *magos*, "wizard, magician." Either of those three words would have gone well with IGIGI, in the senses of: Warlike Igigi, Together with the Igigi, or Magician of the Igigi.

The Igigi, of course, are the spirits who dwell beyond the "zoned ones," beyond the realm of the planets and in the deep vastness of space. Their Underworld counterparts are the Annunaki, with whom they are often confused (perhaps deliberately). To walk among the Igigi is to truly tread among the stars, and this, the last of the Seven Gates, is the Gate to the Igigi, the Gate to the Stars.

There is a scepterlike object in the lower right-hand corner of this glyph, which may indicate to the magician that success has been obtained in combining the forces and maintaining the polarities, for this scepter seems suspiciously similar to the caduceus with its paired "serpents" and central, unifying, disk. Opposite the scepter is a flower of unknown type, perhaps a reference to an ancient drug. Jimson weed and hemp were known to the Chinese alchemists and used by them in their rituals. Although this flower does not bear any resemblance to either of those two plants, it may very well refer to some other plant with properties similar to the aforementioned narcotics.

The three other signs are as mysterious as any of the others we have looked at on the other Gates, and we may pass over them in silence.

The Seal of the Gate

We have before us another symmetrical diagram containing four corner squares, which each support a single letter or glyph. We can make out the Greek letters Sigma, Nu, Lambda, and what may be Upsilon. These letters do not spell anything of themselves, and may be abbreviations for some other forces. The other glyphs defy explanation, save to say that the symbols in the center seem to represent beings floating in space (perhaps the Igigi themselves?), but this is pure fancy on the author's part. A notable feature of this seal is the lack of upper and lower crossbars, which may indicate a sense of disorientation familiar to those who have experienced zero gravity: a sense (or, more accurately, a realization) that there is no "up" or "down" anymore.

Other Information

NINIB called ADAR is a Lord of Hunters and of Strength. He wears a crown of horns (which may explain the crescent shape on the Gate diagram), carries a sword, and wears a lion's skin. As a Hunter, of course, he ventures forth into the wasteland, no-man's-land, the outback of Space. For this reason, "he knows best the territories of the Ancient Ones." His color is Black, his metal Lead, and his essence can be found in the embers of a fire and in "things of death and antiquity."

His number is Four, "as the quarters of the Earth," which may explain the four corner squares of his Seal.

Daoist Analogue

This star, of course, is called "Celestial Gate." Often, the entire Dipper is called the Celestial Gate, for it provides the entrance to the Celestial realms.

It is analogous to another "celestial gate": the one that exists in every human brain. Traveling along the Pace of Yü or the Walking of the Mad Arab, one also travels up the path of energy in the body, the path known as Sushumna to the tantricists, along which the Serpent Goddess Kundalini is raised. The horns of the Sumerian symbol for NINIB becomes the Ram's Head of the Egyptian *Ur-Hekau* atop the serpent staff. All of these symbols are cognate: they share a common origin and a common understanding of how the universe—the *human* universe—really works. The common origin may not have been a specific country or "tribe"; it might have been simply what our bodies and souls tell us when there are no distractions other than Nature herself around us. It might have come from the stars, by way of a messenger like Oannes.

Or it might be the result of intensive scrutiny by engineers of consciousness, magicians and shamans of some lost society that gave all of us yoga, magick, alchemy, divination, and the first broad sketch, the first fuzzy outline, of a Being some call the Monster and others call God.

CHAPTER 15: The Calling

Magickal Evocation

These ceremonies are of a different nature than the ones we have been studying for the Walking. Whereas the Walking is a strictly self-initiatory process, the Calling involves the exercise of powers that have been attained through initiation.

In terms of Western ceremonial magick, the Calling is equivalent to the Golden Dawn's Evocation to Visible Appearance: i.e., it produces the bizarre effect of a visible manifestation of a spiritual entity or force. This can be considered a "self-induced hallucination," of course, but we don't know what *that* means, either. Let us say, for the benefit of skeptics, that the summoning forth of angels, demons, and shades of the dead is, indeed, an exercise in self-induced hallucination. The hallucination is induced by means of a structured ceremony involving specific elements, is controlled, and ends upon command; and, quite often, the hallucination may be shared by others in the vicinity! The hallucinations vary according to the specific elements employed, they often convey information that is of use to those present, and the experience acts as a kind of confirmation that the initiatory program is proceeding smoothly.

The hallucinations are not always visible; they are, at times, audible and not visible, or audible and visible at once.

They can affect any or all of the senses. They may be experienced by others in the vicinity, or not. An often cited requirement of the Golden Dawn organization was that the hallucination be at least partially visible or "sensible" to other members who were present, to judge the capability of the initiate in this regard. Hence, the belief that for an hallucination to be perfect it should be manifest to others as well as to the self-inducer.

The conjuration—or evocation, or summoning—to visible appearance is by far the splashiest aspect of ceremonial magick. The legends that have grown up around the sorcerers and magicians of the past center chiefly on this ability to summon entities to visible appearance. The grimoires are entirely devoted to this aspect of magick, and it is the one responsible for attracting the greater numbers of would-be magicians. Many authors in the past have avoided this area like the plague, simply because they have not performed these rituals and therefore cannot vouch for them or even profess to understand them, or because they have performed these rituals, but poorly, and with little to show for their efforts other than embarrassment. They are afraid to openly discuss this, the most outrageous aspect of Western occultism, for fear of losing their credibility in front of an audience that desires its mysticism served up lukewarm, a bit of trifle after the meal and not a main course.

We are not so timid.

The shamanistic quests of various cultures involve such hallucinations as an integral part of the initiatory experience; indeed, without them the quest has failed. Hallucination is a phenomenon familiar to those who deal with various types of schizophrenia, particularly with paranoid and hebephrenic (disorganized) disorders, and including full-blown psychosis, and for this reason it is generally assumed that all hallucination is a symptom of some sort of mental, emotional *disease*.

In primitive cultures, where the person undergoing hallucinations is treated as someone involved on a spiritual journey—as someone speaking directly to God—the unpleasantness of the experience is more easily accommodated within the social framework. This difference of perception is crucial to an understanding of the entire occult and magickal impulse that is now burgeoning within our otherwise enlightened and scientific Western society. In cutting ourselves off from this source of personal, individual growth and creativity, we have also lost all sense of moral direction. When visions became "hallucinations," and initiation became "schizophrenia," science severed religion from experience with a sharp sword. In the West, it is only within the murky labyrinths of occult ritual and meditation that this rift is healed; in the East, behavior that modern psychology and medical science could only deem psychotic in the extreme is acceptable within a religious, cultic context and is, in fact, encouraged. For instance, we need only venture to India or Southeast Asia to see examples of extraordinary feats of human endurance taking place at the time of various religious festivals; activities that would surely find their perpetrators confined for a period of observation in a mental ward if taking place on the streets of New York or Chicago.

Let's face it: there is no intrinsic danger to society in the visions of an individual who claims to have spoken with the angels. The danger comes when that individual then believes he or she has a mandate from God. In that case, of course, our cherished social institutions are threatened. In the Catholic Church, only the hierarchy can determine whether someone has spoken with God. These are often the people least capable of making that determination.

If, however, we *all* had direct communication with God, there would be no earthly reason for a hierarchy whatsoever, for what would they do? What purpose would they serve? When each person becomes their own priest or priestess, their own Pope, the need for bishops and cardinals dwindles to nothing.

The current situation, however, is one with which Western

society is having a terrible time dealing. The shamanistic experiences continue, unabated, in our streets and within the walls of clinics, but without the context or social framework necessary to enable the individual to interpret the experience or to effect a cure. There is simply no available database of cultural signals we can use to draw the individual along in his or her quest until the goal of unity with the divine Source can be realized and a cure accomplished. Further, we have no need for those people who *have* become cured after a lengthy psychotic or schizophrenic episode. They are social pariahs. We do not wish to hear descriptions of what they have felt or experienced during that period. We want only to forget that such a period ever took place.

In ceremonial magick, however, a framework exists within which we can deal effectively with this experience. It is absolutely necessary, however, that before such an individual undertakes this journey, he or she is provided with the training appropriate to the journey. This involves long periods of study and psychological self-preparedness in which an array of religiocultural icons is schematized in such a way that an entire cosmological pattern is impressed upon the individual's subconscious. These are the tools that will enable the individual to (a) interpret the unconscious material as it becomes conscious, even if in the form of an hallucination, and (b) recognize the experience as a path with an ending, a predictable goal, and not as a chaotic state from which there is no escape, no value to be obtained. Without this period of training, an individual undergoing a nervous breakdown or schizophrenic episode in which hallucinations play a prominent role is adrift on the sea without a rudder, without a paddle, and without the stars above for navigation or guidance. It is also extremely difficult at that time for another individual—magician or shaman though she or he may be, and having gone through the experience already—to rescue the lost person from that stagnant sea. A common language is necessary, a language of symbols that transcends the idiosyncratic and purely personal, a language of symbols that are more or less universal to all human beings: a language that has ordered the visible and invisible universe into comprehensible categories.

A Siberian shaman has a set of these symbols already available in the form of the local culture; there is rarely a need for any prior training in a symbolic language. When an individual shows signs of becoming—what we in the West would call—"deranged," then an existing shaman takes over and cares for the new initiate until the "cure" has been effected (in effect, guiding the experience) and the "deranged" person returns to the village a new shaman.

For us in the West, however, the only coherent symbolic language with any degree of internal consistency is that of the occultist, the magician, the alchemist. The symbology of organized religion is too loaded and weighted to one side to be of much use in a genuine initiatic scenario without the support of a "secret doctrine" to fill in the gaps. (In the Eastern Orthodox Church, the *Philokalia* helped to bridge the gap between organized Christianity and mysticism by instructing its readers in the use of the "Jesus Prayer," a simple mantra that could be used to heighten awareness and bring the chanter to unity within a framework that is suspiciously Eastern and yogic in nature.)

For us, of course, the *Necronomicon* provides just such a framework, as do other forms of ceremonial magick. The framework of the *Necronomicon* is nowhere near as sophisticated or complex as that of the Golden Dawn, for instance, but the elements that do exist are emotionally powerful and readily identifiable. ANU, ENLIL, and ENKI are described at some length, and more information on these figures is available from any library. The Watcher is certainly a potent psychological symbol, as are various of the demonic forces and the Fifty Names of Marduk. The Seven Gates have already been discussed at length, and the process of consciously and deliberately going through them is an educational program that trains the subconscious mind in understanding and manipulating the relevant symbols.

For this reason, the Gates should have been mastered first before undertaking the Calling operations, for the Calling involves opening the last Gate—the Eighth Gate, if you will—which opens up the individual to forces and experiences that lie outside the purview of even traditional occult practice as it is understood in the literature. Nothing—no one—can prepare you for the experiences that await the opening of this Gate; it is, in a sense, entirely personal. For that reason, no one can save you should you rush into these operations without a solid background in the symbol structure.

Of course, we all have these hallucinations all the time: they are called "dreams." The difference between dreams and the hallucinations we are discussing is that the latter take place during full waking consciousness, usually at the instigation of the person having the hallucination, and are consciously controlled, unlike most dreams. The attraction many people felt during the sixties for hallucinogenic drugs such as LSD, mescaline, and psilocybin was due to just this aspect of control. One could summon hallucinations, and one knew that they would eventually end. The problem, of course, was the lack of a symbol system adequate for the interpretation and conscious control of the hallucination as it was taking place. A lack of cultural context.

The Calling

The type of entities that may be summoned are listed in the Book of Calling, the Urilia Text, and the Book of the Fifty Names. These range from the Spirits of the Seven Gates to spirits of the Dead, the Unborn, the Fifty Names themselves, and various monstrous entities. The specifications for these ceremonies are virtually identical to those for the Walking, save that the magic circle is one that is appropriate to the nature of the Calling and for which some examples are given in the Book of Calling. Also, the raiment worn by the summoner is more sophisticated and involves the designs and styles given.

An entreaty that is made several times, however, regards the care that must be taken when "opening the Gate." To the Mad Arab, any act of contacting the Dead, various spirits, or monstrous entities was an act of "opening the Gate": the Gate to the Underworld, which is perceived by him to exist both below the earth and somewhere in space, as we have discussed at length.

To be effective, the Gate should be opened at a specific time. Of course, this time is related to the position of the Dipper and must take place in the evening, preferably between the hours of eleven P.M. and one A.M. As you can see from the instructions regarding the setting up of the ritual site—including the various exorcisms, establishing the "Mandal of Calling," the Four Quarters, etc.—a considerable amount of time is required to perform even the preliminary rituals. These may not be started before sunset, and preferably not until late in the evening. Hence, the actual moment of opening the Gate—the moment when the Invocation to the particular Spirit is being pronounced—may not occur until nearly one A.M. itself, or even later.

The magician should not be concerned with timing the ritual that precisely at first. This is because the experience of performing this ritual is strange enough that concentration on the clock would only serve to hamper the magician from raising the appropriate level of psychic energy necessary to achieve any degree of success. Later, when the basic liturgics of the ritual are mastered, the magician can concentrate on precise times and directions. In other words, the first "ceremonies of Calling" are trial runs; they may seem to be successful, and they may not. Usually they are quite successful, though not in the ways imagined.

For instance, the summoning of a dead relative may not appear to work the first time. There is no sense of the departed having visited the circle, no sign of any hallucination of the senses, no outward manifestations, no psychic phenomena. Yet, later on the next day or during the week, a communication is often received confirming that contact has somehow been made with some aspect of the dead person: a previously unknown letter from that person is suddenly found, a relative dreams of that person or believes to have seen the person's ghost or heard the person's voice, some legal matter concerning the dead person is suddenly resolved, etc.

Necromancy of this type is quite common in every culture; the author himself has been involved on several occasions where the phenomenon of demonic possession was directly related—not to demonic entities as was believed—but to aspects of the deceased "possessing" a relative or neighbor and urging that some action be performed to satisfy some posthumous desire. (This has been especially the case with Asian cultures.) What this actually *means* in terms of a scientific explanation is beyond the scope of this book. The author humbly refers the bewildered reader to the first section of the present volume, the Prolegomena, for a general consideration of how these things happen.

The same is true, of course, not only for works of necromancy (raising the dead, divining the future by means of raising the dead) but also in every type of magickal conjuration, particularly in the early stages of practice. As one becomes more adept, the conjured entities appear more readily. In the case of the shamans we have been discussing, it is not uncommon for the entity so evoked to temporarily possess the shaman, and speak through the shaman (much as our contemporary fad of "channeling" has accustomed us to expect, but with entities that are usually more interesting and with a greater sense of humor). This is true of Siberian and Tibetan forms of shamanism, as well as of Haitian *voudoun* and the other African-American cults that proliferate in Central and South America. Modern Western magicians also employ such mediumship in their rituals, but usually in the form of another person into whom the entity is conjured. In other words, the magician herself or himself is not possessed, but another person is the (voluntary, of course) medium for the ritual. In this way, the magician can more carefully control and direct the experience, and care for the medium's health and safety at the same time.

Interested readers are directed to Crowley's writings concerning rituals involving Victor Neuberg in the North African desert, or those sections of Benvenuto Cellini's autobiography that deal with the famous conjuration in the Colosseum in Rome.

There are a number of basic things to keep in mind when attempting these operations.

1. Once the ritual begins, do not leave the circle until the conjured entity has departed. To do otherwise is to invite disaster: possession by the entity, or worse.

2. Follow the instructions given in the grimoire exactly and to the letter until such time as experience and comprehension permits you to alter the instructions to suit yourself.

3. Copy everything exactly, not changing the style or design of any letter, glyph, or stray line, until you know for certain what was intended and can correct any errors that you may find.

4. If you are accompanied in these rituals, select only companions upon whose judgment and courage you can completely rely, and companions who will follow the instructions and above restrictions as carefully as do you.

5. Do not rush through a ceremony; it will be wasted.

6. Do not arrive at the ritual site fresh from a day's work or from any bother or aggravation, as it will color the performance of the ritual. "Decompress" before beginning. Relax. Approach the ritual site in awe and respect, for it is nothing less than your inner mind made outer, your unconscious identity revealed to your consciousness. These are among the most sacred moments in your life; do not corrupt them with daily worries or cares.

7. Maintain both secrecy and a record. That is, do not speak of these rituals beforehand to anyone, save your selected companions; you endanger yourself and them if you do. Similarly, your companions must be under the same restriction. After the ritual has been performed, a record must be kept of the day, time, and conditions of the rite, the type of rite, and the results. This record can be encoded for your protection, or simply concealed in a safe place. It is not for prying eyes. The record is absolutely necessary; do not neglect to record any ritual. All of this information will become increasingly important to you as your inner journey progresses; the information contained in even the most idle jotting down of the above characteristics is of inestimable value, as you will see years later.

The author wishes to thank everyone who has persevered in studying the *Necronomicon* and in working with it through the years. Occasionally, the demons you evoked have visited him in passing.

The harvest is good, the forests green and welcoming, the sun warm, the moon bright. Do not despair, for the Gate is open to all who approach, in dignity, in truth, and in courage.

Good hunting!

PART THREE

Tables of the Bear

Tables of the Bear

When an observer in the Northern Latitudes faces North—the direction of the Underworld, and where the altar should be placed for the Ceremony of the Walking—the Great Bear hangs from its tail once every day. As a circumpolar constellation, it rotates around the Pole Star (Polaris) once every twenty-four solar hours (approximately: there is roughly a four minute difference between the solar and sidereal clocks, which accounts for the "slide" of the constellation as these four-minute increments accumulate into hours).

However, half of that time the constellation is invisible (i.e., during daylight hours), so that the effective range of its rotation is somewhat limited for our purposes. Although it hangs from its tail once every day, this phenomenon is not *visible* during certain months, and is only visible in the evening (after sunset) during the other months.

This means, however, that the Gate is open once *each day*, whether the constellation is visible or not. The ancient magi who calculated the times of the rites pertinent to the *Necronomicon*, however, would have restricted themselves to the seasons when the phenomenon was visible to them. It would have to be late enough after sundown and long enough before sunrise so that the skies were dark. Those who invoked the Ancient Ones would have waited for the dark of the moon, when the seven principal stars comprising the Dipper asterism were brightest; while those involved in the Ceremony of the Walking worked during the full moon days, as per the rubrics.

As the rituals of the *Necronomicon* are to be performed in the dead of night, we can reasonably expect that only the latest hours of the night are appropriate: from eleven P.M. to one A.M., the "hour" of "rising Yang." With that in mind, the following tables are constructed, correct for those hours at about 40 degrees north latitude. Notations are added to show the desirability of some days over others, and the reasons why. Special astronomical phenomena are taken into account, such as eclipses, full moons, etc. Aspects to the Moon itself are not included (except for important eclipses), as the Moon aspects all planets once every month. For more detailed information, the author recommends a visit to a planetarium or your local library's astronomy section for more precise calculations (if necessary) for your specific latitude and longitude. Also, an astrological almanac or diary or magazine will show various aspects and other indications not given here, which can then be corrected for your latitude and longitude. The phrase "the Great Bear hanging from its tail" should be defined as the moment when Polaris and Arcturus form a line that roughly bisects the horizon in half from North to South, with the Handle of the Dipper between them. The "bowl" of the Dipper, of course, hangs from its handle so that the handle is "on top," as it would be viewed from this angle, from an observer facing North.

A compass or some other form of magnetized needle may be used to position the altar so it faces magnetic north. Of course, this direction is off from celestial "true" north, depending on where you happen to be. In other words, a magnetized needle will not point at the Pole Star or at the Dipper itself (depending on where you are on the planet). The identification of the "south pointing needle" with the Dipper, however, indicates that both of these factors are important. In a way, they are related in that the magnetic poles of the Earth are determined by its wobble on its axis, which is what determines the precession of the equinoxes (the perceived backwards motion of the sun through the signs of the zodiac every year so that the vernal equinox takes place in a different degree of the sidereal zodiac every seventy years or so). What happens is the wobble of the Earth on its axis—like the proverbial spinning top—describes a circle in the heavens, shifting the celestial pole from one star to another, from one constellation to another, and sometimes leaving us with no celestial marker at all, only an "empty" space in the sky.

Magnetic storms—which are massive disturbances in the earth's magnetic field—occur most often at the time of the equinoxes than at any other time of the year. When the equinox coincides with a time of profound sunspot activity, then you have all the ingredients for a magnetic disturbance of great magnitude.

Further, there are daily and hourly variations in the magnetic fields of the Earth that are determined by the Sun and Moon. The first is called the Solar Daily Variation, and is greatest when the Sun crosses the meridian at the place the measurements are being taken: i.e., for our purposes, at the place the Ceremony of the Walking or of the Calling is being performed.

The second is called the Lunar Daily Variation, and is reckoned from the time the Moon crosses the same meridian.

The Solar Daily Variation is more profound during daylight hours, and the Lunar is more profound during the evening hours. Also, the Solar Daily Variation is greater during the summer months than during the winter months. For the Moon, the "lunar seasons" are different, and would probably correspond to the winter months for the Full Moon and the summer months for the New Moon.

This annoying wobble of the oblate spheroid we call Earth occurs with predictable periodicity, so that the star Thuban in Draco was once the Pole Star—circa 3000 B.C.—and will be again, 26,000 years (the precessional period) after the last time it was the Pole Star, or around 23,000 A.D., and so forth.

If the Earth were a perfect sphere, there would be no precession, and celestial north and magnetic north would be identical. It wasn't until the sixteenth century A.D. that formal notice was taken that the magnetic needle did not point to the celestial pole but to a terrestrial pole somewhat at variance with the celestial one. As it is, there is a point on the Earth's equator where a magnetic needle *will* point due north. This point is referred to as "zero declination," and moves at the rate of about one-fifth of a degree of longitude per year. At the moment, it is somewhere in Brazil, having moved here from across the Atlantic, in Africa, and having taken hundreds of years to do so.

There are other places on the Earth's surface where the magnetic field exhibits "peaks" and "valleys," if you will, places of greater and lesser strength that are called isoporic foci.

The point to this discussion is that the Earth's magnetic field is very much a product of a number of intersecting forces: the gravitational pulls of the Sun and the Moon, sunspot activity, and the relation between the Earth's crust, mantle, and cores. Disturbances in the Earth's magnetic field are subject to seasonal, as well as daily, variations and are also subject to the eleven-year sunspot cycle. Further, a meteor shower that happens to pass close enough to either of the magnetic poles can cause widespread electrical outages over large portions of the planet.

Our famous "ozone holes" are, of course, located over the poles, and while they may not be directly related to magnetic influences, they are located at the poles due to the Earth's rotational spin . . . which is a function of gravitational forces, which contribute to magnetic intensities, etc.

Therefore, when the alchemists portrayed the Sun and the Moon in various combinations, indicating an alchemical "wedding," a subtext might very well have been the telluric and magnetic forces at work below the Earth's crust, forces that affect our lives daily in myriad ways, from the tides of the sea to radio and TV reception and, perhaps, to earthquakes and other natural disasters, as well.

For our purposes, celestial "true" north may be the more appropriate placement for the altar than magnetic north, if only because local disturbances in the geomagnetic sphere can confuse the devil out of a compass, and we want the devil to stay put right where it is. On the other hand, the Chinese *feng shui* masters have a lot to teach us about the utility of the magnetic field over and above such a mathematical abstraction as "true" north. Basically, the difference between the two "norths" relate to the difference between earthly, or telluric, energy, and celestial or astral energy. The earth's magnetic field is where the Serpent is to be found in its natural state; the celestial north pole is its destination when awakened. From this, the author believes a careful and serious reader will be able to formulate an independent set of ritual arrangements.

Walpurgisnacht

This famous holiday—familiar to anyone who has seen the original *Dracula* film or read the Bram Stoker novel—is a pagan festival that begins at moonrise on April 30, and is still celebrated in many parts of Europe today as their version of our American Halloween. Actually, these two days are both pagan festivals of great importance, being two of the cross-quarter days that subdivide the zodiacal year; the other two cross-quarter days being August 1 (Lammas) and February 1 (Imbolc). The "quarter" days are, therefore, the two solstices and the two equinoxes. Together, these eight days formed the principal calendar of our European pagan ancestors.

On Walpurgisnacht—May Eve, the 30th day of April—the Great Bear hangs from its tail in the sky at about midnight, and thus is an ideal time for the performance of the *Necronomicon*-related mysteries. But, as this system is not zodiacally structured, there are no days more perfect than others, only times of the day more perfect than other times. The Gate can be opened on any day; a total solar eclipse occurring at a time when the Great Bear hangs from its tail would be a particularly powerful time, as day is changed into night at a time when the Gate is open, and the stars are visible for a short period. During such a time, care must be taken to open and close the Gate before the end of the eclipse. A Gate left open after the eclipse is over is a Gate that cannot be closed until the next day at the same time.

When evoking demons from the *Necronomicon*, twenty-four hours is a *long time*!

During the time when the star Thuban in the constellation Draco was the Pole Star, the vernal equinox took place in Taurus, the sign of the Bull. This constellation is noteworthy for many reasons, not the least of which is that the Pleiades—a cluster of about 250 stars—can be found there, as well as the Hyades and the bright red star, Aldebaran. The resemblance of this odd group of star clusters and individual stars to a bull is not obvious to a modern observer, and were it not for star maps that show how the ancients perceived this constellation to be arranged, it would be virtually impossible to identify it. One cannot help wondering if the constellation was so named for a reason other than its appearance.

Taurus actually appears in the nighttime sky during mid-November, when the Sun is "in" Scorpio, the sign opposite, thus illustrating the polarity between Samhain and Beltane. When the Sun is in Taurus at the time of Beltane, the midnight sky will reveal the sign of Scorpio.

According to the venerable Sumeriologist Samuel Noah Kramer, the ancient Sumerians began their year around April-May, the time of Beltane, even though winter began for them at the time of September-October, and summer in February-March, thus indicating that Beltane, the "cross-quarter" day and a day halfway between the vernal equinox and the summer solstice, was probably the starting point for the year, the Sumerian New Year's Day,[1] a concept that survived in the pagan calendars of Europe. Indeed, even the month of August corresponding to the zodiacal sign Leo was known to the Sumerians as *Lamas* or *Lamash*, which name has survived in the pagan cross-quarter holiday of *Lammas*, August 1.

The cross-quarter days exist at "right angles" to the quarter days of the two solstices and two equinoxes. Using Beltane—the Sumerian New Year's Day—for raising the May Pole seems odd, unless one understands this festival and bizarre calendar to perform as an *anticalendar*, a calendar existing in the *spaces between* the normal, civic one. One can imagine the pagans insisting upon the existence of a completely different world-axis, one that was not dependent on the north-south magnetic or celestial poles, nor on the path of the ecliptic, but one at "right angles" to it.

The equinoxes, after all, take place when the Sun has crossed the equator, which it does twice per year. The equator is at "right angles" to the poles, thus when the Sun intersects the equator, it causes the severe magnetic disturbances we have already discussed. If we posit a different axis, however—a theoretical axis that might be a memory of a shift of magnetic poles that took place many millennia ago—then the insistence on Beltane as a New Year's Day and the erection of a *pole* on this day might be the survival of an ancient civilization that was all but destroyed at the time of the shift of the poles. We could then draw an antiecliptic running at 90 degree angles to the zodiacal belt that would comprise the cross-quarter days as the equinoxes and solstices of time before recorded history.

The recognition by the pagan cults of the holidays taking place at the current equinoxes would reveal an exploitation by them of the very magnetic disturbances we have been discussing, a way "in" to their previous source of power and identity. A Gate.

The Tables

The Tables offered here are divided into two sections, one for the Walking and one for the Calling. The former relies on the age of the Moon throughout the year, and takes into consideration some other astrological phenomena where appropriate. The latter is based on the motions of the Great Bear and the optimum times for opening that Gate.

The Walking section tells you when to begin the various fasts and purifications. Naturally, everything depends on when one begins the rituals. Therefore, no attempt has been made to identify the First Gate, the Second Gate, etc. The times and days are good for all Seven Gates. Also, one need not go through all Seven Gates in seven months. It is possible to stretch this process out over years. For that reason, the Tables here provided are good through December 31, 2018 c.e. The Full Moon is both the day to begin the purification period preceding the Walking and the day to perform the Walking itself, depending upon where you are in your own private schedule. The date—a week before the Walking—that the Meatless Fast begins is noted, as well as the water-only Black Fast three days previous to the Ritual. You will undoubtedly note that the Fast begins on or about the day of the Moon's First Quarter.

The Calling section tells you simply when the Gate—the Gate to the Underworld—is open or able to be opened. The rest is up to you.

Times of Calling—when the Gate to the Underworld is open—are good for northern latitudes of approximately 40 degrees and up. The timing is based on local *standard* time, as calculated from sidereal time, as midnight is midnight wherever you happen to be. Naturally, greater precision can be obtained for your latitude and longitude with reference to your local newspaper. Where Lunar dates and times are given, these are based on ephemeris time (a standard for astrological ephemerides); to obtain local standard time one must add or subtract the hours from ephemeris time (roughly identical to Greenwich Mean Time) for your location. For example, for New York City and for the Eastern seaboard of the United States generally one should subtract five hours from the time shown; however, this information is more readily obtained from your own local newspaper, observatory, or almanac. Remember, the time shown does not take into consideration Daylight Savings Time or other similar corrections; if you are in "Spring forward" time, then the difference between your time and ephemeris time is an hour less.

A note about eclipses:

Since lunar and solar eclipses were always considered important events in the ancient world, their date, time, and characteristics are given here. A total solar eclipse is one where the sun is completely obscured by the moon; a partial solar eclipse is, of course, when only part of the sun's disk is thus obscured. An annular solar eclipse is a total eclipse, but one where the rim of the sun can be seen around the obscuring moon. There is also the annular-total eclipse, which means that the eclipse is total for part of the time and then becomes annular. For the moon, the definitions are similar except there is no annular lunar eclipse but an "appulse" lunar eclipse wherein the moon enters only the penumbra of the earth.

A few astrological events are recorded below where they impact the opening of the Beltane Gate on April 30. As this is not a treatise on astrology, the events are mentioned briefly to give the reader an idea of how to interpret the relation between general astrology and the specific sidereal—or astral—nature of the *Necronomicon*. For more detailed information on the nature of these astrological aspects, one is directed to any good astrological almanac or textbook.

General Observations

From 2006 through 2017, midnight on Beltane will find Pluto—Lord of the Underworld—rising in Capricorn on the eastern horizon. This is considered a very powerful aspect for those working within an occult framework, as it pertains to the darkest occult mysteries. It is also particularly apt for the *Necronomicon* workings, as the Walking pertains to the Underworld and as Pluto, the outermost planet, is assigned to the rulership of Scorpio, the traditional Eighth House or House of Death. Those working with New Aeon formulas will find their work made easier; they should pay particular attention to the Beltane Gate during this period, especially from now until the Mayan date specified for the "end of the world" on December 24, 2012.

The author will set out a few remarks concerning the next few years as a guide for those who wish to delve into the Gates for themselves and adjust their working according to their own cultural or aesthetic sensitivity. Using one's natal birthchart as a touchstone for discovering days and periods of greater individual power and potential is highly recommended. Once again, the times given for the Gate of Calling is local *standard* time: remember to subtract an hour from Daylight Savings Time to find local standard time in the summer months. Thus, if the Gate of Calling is open at 7:23 P.M. in the evening of July 1, your clock might say 8:23 P.M. because of "Spring forward, Fall back."

As mentioned above, the dates and times given for the lunar periods are based on Ephemeris Time (ET). The dates will differ slightly for our American readers, for instance, particularly as many of the lunations occur on or about midnight, Greenwich Mean Time. The best recourse is to obtain a copy of one's local almanac or newspaper for more accurate readings. The purpose of the following Tables is more by way of illustration: it should be quite easy for anyone to draw up their own tables of greater precision, taking into account local eclipse visibility, geomagnetic disturbance, natal chart aspects, etc.

Our Southern Hemisphere readers may be assured that the Gate is opened at the dates and times specified in these Tables, even though the Northern Dipper will not be visible from their latitudes. Remember, it is not the Dipper that is moving: it is the Earth, and the Earth's position is really what is recorded by the times given for the Gate of Calling.

2007

The Beltane Gate opens at 11:30 P.M. on April 30. In addition to Pluto rising in Capricorn, Uranus is conjunct Mars in Pisces and square Jupiter in Sagittarius. Jupiter itself is trine Saturn in Leo. This is a moment of tremendous cosmic tension. A Uranus/Mars conjunction is emblematic of revolution and violence; for an occultist this conjunction can be a source of occult energy and power. The Jupiter/Saturn aspect is beneficent for operations undertaken in seclusion. Jupiter square Uranus/Mars ameliorates somewhat the negative aspect of the conjunction. Thus, an excellent combination for the beginning of a serious spiritual undertaking.

2008

As before, Beltane at 11:30 P.M. sees Pluto rising ever closer in Capricorn. Negative aspects are few on this day, with a Mars/Jupiter opposition and Saturn trine Pluto. A time rife with creative possibilities.

2009

Pluto continues to rise at 11:30 P.M. on Beltane this year. Jupiter and Neptune are conjunct in Aquarius, giving rise to spiritual insights, illumination, and energy for Walking on the stars.

2010

Pluto rising is combined with Saturn in opposition to a Uranus/Jupiter conjunction. The Neptune and Jupiter conjunction of 2009 is replaced by a Uranus and Jupiter conjunction in 2010. Pluto, Neptune, and Uranus are all considered powerful occult symbols, and at Beltane over the past few years Jupiter has aspected one or more of these planets. Jupiter, as the planet of good fortune and the ruler of Sagittarius, is a beneficial influence upon these darker forces, enabling the magician to control the uncontrollable.

2011

Pluto rising still, this year's Beltane sees a powerful conjunction of the Dark Moon with Mercury, Venus, Mars, and Jupiter in Aries, in opposition to Saturn in Libra. While there are no especially powerful aspects to Pluto, Neptune, or Uranus, this "stellium" of planets in Aries in opposition to Saturn in Libra—with Pluto still in Capricorn—involves six of the seven "planets" that represent the seven Gates, and as such should be an auspicious moment for beginning the Walking or for completing it if it has already been begun.

2012

As before, Pluto is still rising at the moment the Beltane Gate opens at 11:30 P.M. At the same time, the Moon is conjunct Mars in Virgo and in opposition to Neptune in Pisces. According to our understanding of the Mayan calendar, the world ends on December 24 this year. While anything is possible, of course, care should still be taken to observe the progress of all occult-oriented phenomena this year as it progresses towards the winter solstice. Great care should be taken to ensure that the Gate is not left open beyond its time.

2013

This year, Pluto is precisely conjunct the Ascendant—the eastern horizon—at the time of the opening of the Beltane

Gate. In addition, it is square Uranus in Aries. The Gate will swing open powerfully on this date, and with Saturn conjunct the Midheaven at the same time, there is a possibility that unwanted entities will slither through the Gate at this time. As last year, great care must be taken to follow the rituals scrupulously and with attention to detail.

2014

At the time of the opening of the Beltane Gate, Pluto is in Capricorn but below the horizon; however, it is in opposition to Jupiter in Cancer, and square Mars in Libra and Uranus in Aries. This Cross involving the four Cardinal signs is another powerful aspect for the magician, as the quadrature (as it is sometimes called) is believed to be an indicator of earthly upheaval, such as earthquakes. In this case, the Cross takes place at the New Moon and during an annular solar eclipse. Eclipses are also thought to be predictors of earthly cataclysms, so this combination of events, timed with the opening of the Beltane Gate, demands great caution even as it is a potential source of power.

2015

The astrological year begins on March 20 with a New Moon and a total solar eclipse, followed by a total lunar eclipse on April 4, thus setting the stage for the Beltane Gate on April 30 with Pluto still square Uranus, these two Dark Lords fighting over control of the Gate to the Underworld.

2016

The cosmic forces seem to be at rest for the opening of the Beltane Gate on April 30. It is a welcome respite from the intense stresses of the past seven years, and the Gate opens a bit more quietly than before.

2017

With Pluto in Capricorn opposite to the Moon (in its native sign of Cancer) and square Jupiter in Libra, we have a T-square configuration involving the Cardinal signs on April 30, which may indicate once again a number of telluric stresses. While traditional astrology speaks of the square aspect as being negative, a more modern interpretation prefers to think of it as "powerful" and full of potential energy to be tapped. The magician thinks this way as well.

2018

Beltane begins with Pluto in Capricorn conjunct Mars. The contact between these two important malefics at the time of the opening of the Beltane Gate stresses the importance of the purification ceremonies and the careful maintenance of the Watcher, for this is a dangerous moment when things can go terribly wrong. Strict adherence to the requirements of the ritual is always mandated, but in this case to ignore the strictures of the rubrics is to invite disaster. Great care should be taken when approaching the Gate at this time.

This should give the reader an idea of how to gauge the probable effects of her or his rituals for the remaining years and for days that have not been discussed. Again, as the ancient Chinese understood, it is necessary to align the Table of Earth with the Table of Heaven, analogously, it is necessary to align one's personal abilities and powers with those that obtain in the macrocosm. A careful utilization of one's own natal astrological data with the Tables appended here will ensure greater success for the apprentice magician.

While the Gates are ready to be opened on the dates and times listed, the appropriate lock is within the magician's own personal grasp and cannot be "given" or "loaned" to you by another. That is what protects magick from the abuses entertained by organized religions and mystical movements of all types: no "benevolent monarch" (an oxymoron if ever there was one) can bestow the key, neither through a laying-on of hands nor by selling a diploma. The most anyone can do is give you a set of instructions. The rest is up to you.

Tables of the Bear

Date	Day of Walking	Day of the Moon	Time of Calling	Notes (*Times in Ephemeris Time*)
2006 31-Oct		10	11:28 AM	Samhain
1-Nov		11	11:24	
2-Nov	Black Fast	12	11:20	
3-Nov		13	11:16	
4-Nov		14	11:12	
5-Nov	Full Moon	15	11:08	
6-Nov		16	11:04	
7-Nov		17	11:00	
8-Nov		18	10:57	
9-Nov		19	10:53	
10-Nov		20	10:49	
11-Nov		21	10:45	
12-Nov		22	10:41	
13-Nov		23	10:37	
14-Nov		24	10:33	
15-Nov		25	10:29	
16-Nov		26	10:25	
17-Nov		27	10:21	
18-Nov		28	10:17	

19-Nov		29	10:13
20-Nov	New Moon	0	10:09
21-Nov		1	10:05
22-Nov		2	10:01
23-Nov		3	9:57
24-Nov		4	9:54
25-Nov		5	9:50
26-Nov		6	9:46
27-Nov		7	9:42
28-Nov		8	9:38
29-Nov	Fast Begins	9	9:34
30-Nov		10	9:30
1-Dec		11	9:26
2-Dec	Black Fast	12	9:22
3-Dec		13	9:18
4-Dec		14	9:14
5-Dec	Full Moon	15	9:10
6-Dec		16	9:06
7-Dec		17	9:02
8-Dec		18	8:58
9-Dec		19	8:54
10-Dec		20	8:50
11-Dec		21	8:46
12-Dec		22	8:42

Date	Day of Walking	Day of the Moon	Time of Calling	Notes *(Times in Ephemeris Time)*
13-Dec		23	8:38 AM	
14-Dec		24	8:34	
15-Dec		25	8:30	
16-Dec		26	8:26	
17-Dec		27	8:22	
18-Dec		28	8:18	
19-Dec		29	8:14	
20-Dec	New Moon	0	8:10	
21-Dec		1	8:06	
22-Dec		2	8:02	Winter Solstice 7:23 PM
23-Dec		3	7:58	
24-Dec		4	7:54	
25-Dec		5	7:51	
26-Dec		6	7:47	
27-Dec		7	7:43	
28-Dec	Fast Begins	8	7:39	
29-Dec		9	7:35	
30-Dec		10	7:31	
31-Dec	Black Fast	11	7:27	
2007 1-Jan		12	7:23	

2-Jan		13	7:19
3-Jan	Full Moon	14	7:15
4-Jan		15	7:11
5-Jan		16	7:07
6-Jan		17	7:03
7-Jan		18	6:59
8-Jan		19	6:55
9-Jan		20	6:51
10-Jan		21	6:47
11-Jan		22	6:44
12-Jan		23	6:40
13-Jan		24	6:36
14-Jan		25	6:32
15-Jan		26	6:28
16-Jan		27	6:24
17-Jan		28	6:20
18-Jan		29	6:16
19-Jan	New Moon	0	6:12
20-Jan		1	6:08
21-Jan		2	6:04
22-Jan		3	6:00
23-Jan		4	5:56
24-Jan		5	5:52
25-Jan		6	5:48

Date	Day of Walking	Day of the Moon	Time of Calling	Notes *(Times in Ephemeris Time)*
26-Jan		7	5:45 AM	
27-Jan	Fast Begins	8	5:41	
28-Jan		9	5:37	
29-Jan		10	5:33	
30-Jan	Black Fast	11	5:29	
31-Jan		12	5:25	Imbolc
1-Feb		13	5:21	
2-Feb	Full Moon	14	5:17	
3-Feb		15	5:13	
4-Feb		16	5:09	
5-Feb		17	5:05	
6-Feb		18	5:01	
7-Feb		19	4:57	
8-Feb		20	4:53	
9-Feb		21	4:49	
10-Feb		22	4:45	
11-Feb		23	4:42	
12-Feb		24	4:38	
13-Feb		25	4:34	
14-Feb		26	4:30	
15-Feb		27	4:26	

16-Feb		28	4:22	
17-Feb	New Moon	0	4:18	
18-Feb		1	4:14	
19-Feb		2	4:10	
20-Feb		3	4:06	
21-Feb		4	4:02	
22-Feb		5	3:58	
23-Feb		6	3:54	
24-Feb		7	3:50	
25-Feb	Fast Begins	8	3:46	
26-Feb		9	3:42	
27-Feb		10	3:38	
28-Feb	Black Fast	11	3:34	
1-Mar		12	3:30	
2-Mar		13	3:26	
3-Mar	Full Moon	14	3:22	Total Lunar Eclipse 6:22 PM
4-Mar		15	3:18	
5-Mar		16	3:14	
6-Mar		17	3:10	
7-Mar		18	3:06	
8-Mar		19	3:02	
9-Mar		20	2:58	
10-Mar		21	2:54	
11-Mar		22	2:50	

Date	Day of Walking	Day of the Moon	Time of Calling	Notes *(Times in Ephemeris Time)*
12-Mar		23	2:46 AM	
13-Mar		24	2:42	
14-Mar		25	2:38	
15-Mar		26	2:34	
16-Mar		27	2:30	
17-Mar		28	2:26	
18-Mar	New Moon	0	2:22	Partial Solar Eclipse 9:32:57 PM
19-Mar		1	2:18	
20-Mar		2	2:14	Vernal Equinox 7:09 PM
21-Mar		3	2:10	
22-Mar		4	2:06	
23-Mar		5	2:02	
24-Mar		6	1:58	
25-Mar		7	1:54	
26-Mar		8	1:50	
27-Mar	Fast Begins	9	1:46	
28-Mar		10	1:42	
29-Mar		11	1:38	
30-Mar	Black Fast	12	1:34	
31-Mar		13	1:30	
1-Apr		14	1:26	

2-Apr	Full Moon	15	1:22
3-Apr		16	1:18
4-Apr		17	1:14
5-Apr		18	1:10
6-Apr		19	1:06
7-Apr		20	1:02
8-Apr		21	12:58
9-Apr		22	12:54
10-Apr		23	12:50
11-Apr		24	12:46
12-Apr		25	12:42
13-Apr		26	12:38
14-Apr		27	12:34
15-Apr		28	12:30
16-Apr		29	12:25
17-Apr	New Moon	0	12:21
18-Apr		1	12:17
19-Apr		2	12:13
20-Apr		3	12:09
21-Apr		4	12:05
22-Apr		5	12:01 AM
23-Apr		6	11:57 PM
24-Apr		7	11:53
25-Apr		8	11:49

Date	Day of Walking	Day of the Moon	Time of Calling	Notes *(Times in Ephemeris Time)*
26-Apr	Fast Begins	9	11:45 PM	
27-Apr		10	11:41	
28-Apr		11	11:37	
29-Apr	Black Fast	12	11:33	
30-Apr		13	11:30	Beltane
1-May		14	11:26	
2-May	Full Moon	15	11:22	
3-May		16	11:18	
4-May		17	11:14	
5-May		18	11:10	
6-May		19	11:06	
7-May		20	11:02	
8-May		21	10:58	
9-May		22	10:54	
10-May		23	10:50	
11-May		24	10:46	
12-May		25	10:42	
13-May		26	10:38	
14-May		27	10:34	
15-May		28	10:31	
16-May	New Moon	0	10:27	

17-May		1	10:23
18-May		2	10:19
19-May		3	10:15
20-May		4	10:11
21-May		5	10:07
22-May		6	10:03
23-May		7	9:59
24-May		8	9:55
25-May		9	9:51
26-May	Fast Begins	10	9:47
27-May		11	9:43
28-May		12	9:39
29-May	Black Fast	13	9:35
30-May		14	9:31
31-May		15	9:28
1-Jun	Full Moon	16	9:24
2-Jun		17	9:20
3-Jun		18	9:16
4-Jun		19	9:12
5-Jun		20	9:08
6-Jun		21	9:04
7-Jun		22	9:00
8-Jun		23	8:56
9-Jun		24	8:52

Date	Day of Walking	Day of the Moon	Time of Calling	Notes (*Times in Ephemeris Time*)
10-Jun		25	8:48 PM	
11-Jun		26	8:44	
12-Jun		27	8:40	
13-Jun		28	8:36	
14-Jun		29	8:32	
15-Jun	New Moon	0	8:28	
16-Jun		1	8:25	
17-Jun		2	8:21	
18-Jun		3	8:17	
19-Jun		4	8:13	
20-Jun		5	8:09	
21-Jun		6	8:05	Summer Solstice 6:08 PM
22-Jun		7	8:01	
23-Jun		8	7:57	
24-Jun	Fast Begins	9	7:53	
25-Jun		10	7:49	
26-Jun		11	7:45	
27-Jun	Black Fast	12	7:41	
28-Jun		13	7:37	
29-Jun		14	7:33	
30-Jun	Full Moon	15	7:29	

1-Jul		16	7:25	
2-Jul		17	7:22	
3-Jul		18	7:18	
4-Jul		19	7:14	
5-Jul		20	7:10	
6-Jul		21	7:06	
7-Jul		22	7:02	
8-Jul		23	6:58	
9-Jul		24	6:54	
10-Jul		25	6:50	
11-Jul		26	6:46	
12-Jul		27	6:42	
13-Jul		28	6:38	
14-Jul	New Moon	0	6:34	
15-Jul		1	6:30	
16-Jul		2	6:26	
17-Jul		3	6:22	
18-Jul		4	6:19	
19-Jul		5	6:15	
20-Jul		6	6:11	
21-Jul		7	6:07	
22-Jul		8	6:03	
23-Jul		9	5:59	
24-Jul	Fast Begins	10	5:55	

Date	Day of Walking	Day of the Moon	Time of Calling	Notes *(Times in Ephemeris Time)*
25-Jul		11	5:51 PM	
26-Jul		12	5:47	
27-Jul	Black Fast	13	5:43	
28-Jul		14	5:39	
29-Jul		15	5:35	
30-Jul	Full Moon	16	5:31	
31-Jul		17	5:27	
1-Aug		18	5:23	Lammas
2-Aug		19	5:19	
3-Aug		20	5:18	
4-Aug		21	5:14	
5-Aug		22	5:10	
6-Aug		23	5:06	
7-Aug		24	5:02	
8-Aug		25	4:58	
9-Aug		26	4:54	
10-Aug		27	4:50	
11-Aug		28	4:46	
12-Aug	New Moon	0	4:42	
13-Aug		1	4:38	
14-Aug		2	4:34	

15-Aug		3	4:30	
16-Aug		4	4:26	
17-Aug		5	4:22	
18-Aug		6	4:18	
19-Aug		7	4:16	
20-Aug		8	4:12	
21-Aug		9	4:08	
22-Aug	Fast Begins	10	4:04	
23-Aug		11	4:00	
24-Aug		12	3:56	
25-Aug	Black Fast	13	3:52	
26-Aug		14	3:48	
27-Aug		15	3:44	
28-Aug	Full Moon	16	3:40	Total Lunar Eclipse 10:38 AM
29-Aug		17	3:36	
30-Aug		18	3:32	
31-Aug		19	3:28	
1-Sep		20	3:24	
2-Sep		21	3:20	
3-Sep		22	3:16	
4-Sep		23	3:12	
5-Sep		24	3:09	
6-Sep		25	3:05	
7-Sep		26	3:01	

Date	Day of Walking	Day of the Moon	Time of Calling	Notes (Times in Ephemeris Time)
8-Sep		27	2:57 PM	
9-Sep		28	2:53	
10-Sep		29	2:49	
11-Sep	New Moon	0	2:45	Partial Solar Eclipse 12:32 PM
12-Sep		1	2:41	
13-Sep		2	2:37	
14-Sep		3	2:33	
15-Sep		4	2:29	
16-Sep		5	2:25	
17-Sep		6	2:21	
18-Sep		7	2:17	
19-Sep		8	2:13	
20-Sep	Fast Begins	9	2:09	
21-Sep		10	2:06	
22-Sep		11	2:02	
23-Sep	Black Fast	12	1:58	Vernal Equinox 7:16 PM
24-Sep		13	1:54	
25-Sep		14	1:50	
26-Sep	Full Moon	15	1:46	
27-Sep		16	1:42	
28-Sep		17	1:38	

29-Sep		18	1:34	
30-Sep		19	1:30	
1-Oct		20	1:26	
2-Oct		21	1:22	
3-Oct		22	1:18	
4-Oct		23	1:14	
5-Oct		24	1:10	
6-Oct		25	1:06	
7-Oct		26	1:03	
8-Oct		27	12:59	
9-Oct		28	12:55	
10-Oct		29	12:51	
11-Oct	New Moon	0	12:47	
12-Oct		1	12:43	
13-Oct		2	12:39	
14-Oct		3	12:35	
15-Oct		4	12:31	
16-Oct		5	12:27	
17-Oct		6	12:23	
18-Oct		7	12:19	
19-Oct		8	12:15	
20-Oct	Fast Begins	9	12:11	
21-Oct		10	12:07	
22-Oct		11	12:03	

Date	Day of Walking	Day of the Moon	Time of Calling	Notes *(Times in Ephemeris Time)*
23-Oct	Black Fast	12	12:00 PM	
24-Oct		13	11:56 AM	
25-Oct		14	11:52	
26-Oct	Full Moon	15	11:48	
27-Oct		16	11:44	
28-Oct		17	11:40	
29-Oct		18	11:36	
30-Oct		19	11:32	
31-Oct		20	11:28	Samhain
1-Nov		21	11:24	
2-Nov		22	11:20	
3-Nov		23	11:16	
4-Nov		24	11:12	
5-Nov		25	11:08	
6-Nov		26	11:04	
7-Nov		27	11:00	
8-Nov		28	10:57	
9-Nov	New Moon	0	10:53	
10-Nov		1	10:49	
11-Nov		2	10:45	
12-Nov		3	10:41	

13-Nov		4	10:37
14-Nov		5	10:33
15-Nov		6	10:29
16-Nov		7	10:25
17-Nov		8	10:21
18-Nov	Fast Begins	9	10:17
19-Nov		10	10:13
20-Nov		11	10:09
21-Nov	Black Fast	12	10:05
22-Nov		13	10:01
23-Nov		14	9:57
24-Nov	Full Moon	15	9:54
25-Nov		16	9:50
26-Nov		17	9:46
27-Nov		18	9:42
28-Nov		19	9:38
29-Nov		20	9:34
30-Nov		21	9:30
1-Dec		22	9:26
2-Dec		23	9:22
3-Dec		24	9:18
4-Dec		25	9:14
5-Dec		26	9:10
6-Dec		27	9:06

Date	Day of Walking	Day of the Moon	Time of Calling	Notes *(Times in Ephemeris Time)*
7-Dec		28	9:02 AM	
8-Dec		29	8:58	
9-Dec	New Moon	0	8:54	
10-Dec		1	8:50	
11-Dec		2	8:46	
12-Dec		3	8:42	
13-Dec		4	8:38	
14-Dec		5	8:34	
15-Dec		6	8:30	
16-Dec		7	8:26	
17-Dec		8	8:22	
18-Dec	Fast Begins	9	8:18	
19-Dec		10	8:14	
20-Dec		11	8:10	
21-Dec	Black Fast	12	8:06	
22-Dec		13	8:02	Winter Solstice 6:09 AM
23-Dec		14	7:58	
24-Dec	Full Moon	15	7:54	
25-Dec		16	7:51	
26-Dec		17	7:47	
27-Dec		18	7:43	

	28-Dec		19	7:39
	29-Dec		20	7:35
	30-Dec		21	7:31
	31-Dec		22	7:27
2008	1-Jan		23	7:23
	2-Jan		24	7:19
	3-Jan		25	7:15
	4-Jan		26	7:11
	5-Jan		27	7:07
	6-Jan		28	7:03
	7-Jan		29	6:59
	8-Jan	New Moon	0	6:55
	9-Jan		1	6:51
	10-Jan		2	6:47
	11-Jan		3	6:44
	12-Jan		4	6:40
	13-Jan		5	6:36
	14-Jan		6	6:32
	15-Jan		7	6:28
	16-Jan	Fast Begins	8	6:24
	17-Jan		9	6:20
	18-Jan		10	6:16
	19-Jan	Black Fast	11	6:12

Date	Day of Walking	Day of the Moon	Time of Calling	Notes *(Times in Ephemeris Time)*
20-Jan		12	6:08 AM	
21-Jan		13	6:04	
22-Jan	Full Moon	14	6:00	
23-Jan		15	5:56	
24-Jan		16	5:52	
25-Jan		17	5:48	
26-Jan		18	5:45	
27-Jan		19	5:41	
28-Jan		20	5:37	
29-Jan		21	5:33	
30-Jan		22	5:29	
31-Jan		23	5:25	Imbolc
1-Feb		24	5:21	
2-Feb		25	5:17	
3-Feb		26	5:13	
4-Feb		27	5:09	
5-Feb		28	5:05	
6-Feb		29	5:01	
7-Feb	New Moon	0	4:57	Annular Solar Eclipse 3:56 AM
8-Feb		1	4:53	
9-Feb		2	4:49	

10-Feb		3	4:45	
11-Feb		4	4:42	
12-Feb		5	4:38	
13-Feb		6	4:34	
14-Feb		7	4:30	
15-Feb	Fast Begins	8	4:26	
16-Feb		9	4:22	
17-Feb		10	4:18	
18-Feb	Black Fast	11	4:14	
19-Feb		12	4:10	
20-Feb		13	4:06	
21-Feb	Full Moon	14	4:02	Total Solar Eclipse 3:27 AM
22-Feb		15	3:58	
23-Feb		16	3:54	
24-Feb		17	3:50	
25-Feb		18	3:46	
26-Feb		19	3:42	
27-Feb		20	3:38	
28-Feb		21	3:34	
29-Feb		22	3:30	
1-Mar		23	3:26	
2-Mar		24	3:22	
3-Mar		25	3:18	
4-Mar		26	3:14	

Date	Day of Walking	Day of the Moon	Time of Calling	Notes *(Times in Ephemeris Time)*
5-Mar		27	3:10 AM	
6-Mar		28	3:06	
7-Mar	New Moon	0	3:02	
8-Mar		1	2:58	
9-Mar		2	2:54	
10-Mar		3	2:50	
11-Mar		4	2:46	
12-Mar		5	2:42	
13-Mar		6	2:38	
14-Mar		7	2:35	
15-Mar	Fast Begins	8	2:32	
16-Mar		9	2:30	
17-Mar		10	2:26	
18-Mar	Black Fast	11	2:22	
19-Mar		12	2:18	
20-Mar		13	2:14	Vernal Equinox 5:49 AM
21-Mar	Full Moon	14	2:10	
22-Mar		15	2:06	
23-Mar		16	2:02	
24-Mar		17	1:58	
25-Mar		18	1:54	

26-Mar		19	1:50
27-Mar		20	1:46
28-Mar		21	1:42
29-Mar		22	1:38
30-Mar		23	1:34
31-Mar		24	1:30
1-Apr		25	1:26
2-Apr		26	1:22
3-Apr		27	1:18
4-Apr		28	1:14
5-Apr		29	1:10
6-Apr	New Moon	0	1:06
7-Apr		1	1:02
8-Apr		2	12:58
9-Apr		3	12:54
10-Apr		4	12:50
11-Apr		5	12:46
12-Apr		6	12:42
13-Apr		7	12:38
14-Apr	Fast Begins	8	12:34
15-Apr		9	12:30
16-Apr		10	12:25
17-Apr	Black Fast	11	12:21
18-Apr		12	12:17

Date	Day of Walking	Day of the Moon	Time of Calling	Notes *(Times in Ephemeris Time)*
19-Apr		13	12:13 AM	
20-Apr	Full Moon	14	12:09	
21-Apr		15	12:05	
22-Apr		16	12:01 AM	
23-Apr		17	11:57 PM	
24-Apr		18	11:53	
25-Apr		19	11:49	
26-Apr		20	11:45	
27-Apr		21	11:41	
28-Apr		22	11:37	
29-Apr		23	11:33	
30-Apr		24	11:30	Beltane
1-May		25	11:26	
2-May		26	11:22	
3-May		27	11:18	
4-May		28	11:14	
5-May	New Moon	0	11:10	
6-May		1	11:06	
7-May		2	11:02	
8-May		3	10:58	
9-May		4	10:54	

10-May		5	10:50	
11-May		6	10:46	
12-May		7	10:42	
13-May		8	10:38	
14-May	Fast Begins	9	10:34	
15-May		10	10:31	
16-May		11	10:27	
17-May	Black Fast	12	10:23	
18-May		13	10:19	
19-May		14	10:15	
20-May	Full Moon	15	10:11	
21-May		16	10:07	
22-May		17	10:03	
23-May		18	9:59	
24-May		19	9:55	
25-May		20	9:51	
26-May		21	9:47	
27-May		22	9:43	
28-May		23	9:39	
29-May		24	9:35	
30-May		25	9:31	
31-May		26	9:28	
1-Jun		27	9:24	
2-Jun		28	9:20	

Date	Day of Walking	Day of the Moon	Time of Calling	Notes (*Times in Ephemeris Time*)
3-Jun	New Moon	0	9:16 PM	
4-Jun		1	9:12	
5-Jun		2	9:08	
6-Jun		3	9:04	
7-Jun		4	9:00	
8-Jun		5	8:56	
9-Jun		6	8:52	
10-Jun		7	8:48	
11-Jun		8	8:44	
12-Jun	Fast Begins	9	8:40	
13-Jun		10	8:36	
14-Jun		11	8:32	
15-Jun	Black Fast	12	8:28	
16-Jun		13	8:25	
17-Jun		14	8:21	
18-Jun	Full Moon	15	8:17	
19-Jun		16	8:13	
20-Jun		17	8:09	
21-Jun		18	8:05	Summer Solstice 12:00 AM
22-Jun		19	8:01	
23-Jun		20	7:57	

24-Jun		21	7:53
25-Jun		22	7:49
26-Jun		23	7:45
27-Jun		24	7:41
28-Jun		25	7:37
29-Jun		26	7:33
30-Jun		27	7:29
1-Jul		28	7:25
2-Jul		29	7:22
3-Jul	New Moon	0	7:18
4-Jul		1	7:14
5-Jul		2	7:10
6-Jul		3	7:06
7-Jul		4	7:02
8-Jul		5	6:58
9-Jul		6	6:54
10-Jul		7	6:50
11-Jul		8	6:46
12-Jul	Fast Begins	9	6:42
13-Jul		10	6:38
14-Jul		11	6:34
15-Jul	Black Fast	12	6:30
16-Jul		13	6:26
17-Jul		14	6:22

Date	Day of Walking	Day of the Moon	Time of Calling	Notes *(Times in Ephemeris Time)*
18-Jul	Full Moon	15	6:19 PM	
19-Jul		16	6:15	
20-Jul		17	6:11	
21-Jul		18	6:07	
22-Jul		19	6:03	
23-Jul		20	5:59	
24-Jul		21	5:55	
25-Jul		22	5:51	
26-Jul		23	5:47	
27-Jul		24	5:43	
28-Jul		25	5:39	
29-Jul		26	5:35	
30-Jul		27	5:31	
31-Jul		28	5:27	
1-Aug	New Moon	0	5:23	Lammas/Total Solar Eclipse 10:22 AM
2-Aug		1	5:19	
3-Aug		2	5:18	
4-Aug		3	5:14	
5-Aug		4	5:10	
6-Aug		5	5:06	
7-Aug		6	5:02	

8-Aug		7	4:58	
9-Aug		8	4:54	
10-Aug	Fast Begins	9	4:50	
11-Aug		10	4:46	
12-Aug		11	4:42	
13-Aug	Black Fast	12	4:38	
14-Aug		13	4:34	
15-Aug		14	4:30	
16-Aug	Full Moon	15	4:26	
17-Aug		16	4:22	
18-Aug		17	4:18	
19-Aug		18	4:16	
20-Aug		19	4:12	
21-Aug		20	4:08	
22-Aug		21	4:04	
23-Aug		22	4:00	
24-Aug		23	3:56	
25-Aug		24	3:52	
26-Aug		25	3:48	
27-Aug		26	3:44	
28-Aug		27	3:40	
29-Aug		28	3:36	
30-Aug	New Moon	0	3:32	
31-Aug		1	3:28	

Date	Day of Walking	Day of the Moon	Time of Calling	Notes *(Times in Ephemeris Time)*
1-Sep		2	3:24 PM	
2-Sep		3	3:20	
3-Sep		4	3:16	
4-Sep		5	3:12	
5-Sep		6	3:09	
6-Sep		7	3:05	
7-Sep		8	3:01	
8-Sep		9	2:57	
9-Sep	Fast Begins	10	2:53	
10-Sep		11	2:49	
11-Sep		12	2:45	
12-Sep	Black Fast	13	2:41	
13-Sep		14	2:37	
14-Sep		15	2:33	
15-Sep	Full Moon	16	2:29	Partial Lunar Eclipse 9:11 PM
16-Sep		17	2:25	
17-Sep		18	2:21	
18-Sep		19	2:17	
19-Sep		20	2:13	
20-Sep		21	2:09	
21-Sep		22	2:06	

22-Sep		23	2:02	Vernal Equinox 3:46 PM
23-Sep		24	1:58	
24-Sep		25	1:54	
25-Sep		26	1:50	
26-Sep		27	1:46	
27-Sep		28	1:42	
28-Sep		29	1:38	
29-Sep	New Moon	0	1:34	
30-Sep		1	1:30	
1-Oct		2	1:26	
2-Oct		3	1:22	
3-Oct		4	1:18	
4-Oct		5	1:14	
5-Oct		6	1:10	
6-Oct		7	1:06	
7-Oct		8	1:03	
8-Oct	Fast Begins	9	12:59	
9-Oct		10	12:55	
10-Oct		11	12:51	
11-Oct	Black Fast	12	12:47	
12-Oct		13	12:43	
13-Oct		14	12:39	
14-Oct	Full Moon	15	12:35	
15-Oct		16	12:31	

Date	Day of Walking	Day of the Moon	Time of Calling	Notes (*Times in Ephemeris Time*)
16-Oct		17	12:27 PM	
17-Oct		18	12:23	
18-Oct		19	12:19	
19-Oct		20	12:15	
20-Oct		21	12:11	
21-Oct		22	12:07	
22-Oct		23	12:03	
23-Oct		24	12:00 PM	
24-Oct		25	11:56 AM	
25-Oct		26	11:52	
26-Oct		27	11:48	
27-Oct		28	11:44	
28-Oct	New Moon	0	11:40	
29-Oct		1	11:36	
30-Oct		2	11:32	
31-Oct		3	11:28	Samhain
1-Nov		4	11:24	
2-Nov		5	11:20	
3-Nov		6	11:16	
4-Nov		7	11:12	
5-Nov		8	11:08	

6-Nov		9	11:04
7-Nov	Fast Begins	10	11:00
8-Nov		11	10:57
9-Nov		12	10:53
10-Nov	Black Fast	13	10:49
11-Nov		14	10:45
12-Nov		15	10:41
13-Nov	Full Moon	16	10:37
14-Nov		17	10:33
15-Nov		18	10:29
16-Nov		19	10:25
17-Nov		20	10:21
18-Nov		21	10:17
19-Nov		22	10:13
20-Nov		23	10:09
21-Nov		24	10:05
22-Nov		25	10:01
23-Nov		26	9:57
24-Nov		27	9:54
25-Nov		28	9:50
26-Nov		29	9:46
27-Nov	New Moon	0	9:42
28-Nov		1	9:38
29-Nov		2	9:34

Date	Day of Walking	Day of the Moon	Time of Calling	Notes (*Times in Ephemeris Time*)
30-Nov		3	9:30 AM	
1-Dec		4	9:26	
2-Dec		5	9:22	
3-Dec		6	9:18	
4-Dec		7	9:14	
5-Dec		8	9:10	
6-Dec	Fast Begins	9	9:06	
7-Dec		10	9:02	
8-Dec		11	8:58	
9-Dec	Black Fast	12	8:54	
10-Dec		13	8:50	
11-Dec		14	8:46	
12-Dec	Full Moon	15	8:42	
13-Dec		16	8:38	
14-Dec		17	8:34	
15-Dec		18	8:30	
16-Dec		19	8:26	
17-Dec		20	8:22	
18-Dec		21	8:18	
19-Dec		22	8:14	
20-Dec		23	8:10	

	21-Dec		24	8:06	Winter Solstice 12:05 PM
	22-Dec		25	8:02	
	23-Dec		26	7:58	
	24-Dec		27	7:54	
	25-Dec		28	7:51	
	26-Dec		29	7:47	
	27-Dec	New Moon	0	7:43	
	28-Dec		1	7:39	
	29-Dec		2	7:35	
	30-Dec		3	7:31	
	31-Dec		4	7:27	
2009	1-Jan		5	7:23	
	2-Jan		6	7:19	
	3-Jan		7	7:15	
	4-Jan		8	7:11	
	5-Jan	Fast Begins	9	7:07	
	6-Jan		10	7:03	
	7-Jan		11	6:59	
	8-Jan	Black Fast	12	6:55	
	9-Jan		13	6:51	
	10-Jan		14	6:47	
	11-Jan	Full Moon	15	6:44	
	12-Jan		16	6:40	

Date	Day of Walking	Day of the Moon	Time of Calling	Notes (*Times in Ephemeris Time*)
13-Jan		17	6:36 AM	
14-Jan		18	6:32	
15-Jan		19	6:28	
16-Jan		20	6:24	
17-Jan		21	6:20	
18-Jan		22	6:16	
19-Jan		23	6:12	
20-Jan		24	6:08	
21-Jan		25	6:04	
22-Jan		26	6:00	
23-Jan		27	5:56	
24-Jan		28	5:52	
25-Jan		29	5:48	
26-Jan	New Moon	0	5:45	Annular Solar Eclipse 7:59 AM
27-Jan		1	5:41	
28-Jan		2	5:37	
29-Jan		3	5:33	
30-Jan		4	5:29	
31-Jan		5	5:25	Imbolc
1-Feb		6	5:21	
2-Feb		7	5:17	

3-Feb	Fast Begins	8	5:13	
4-Feb		9	5:09	
5-Feb		10	5:05	
6-Feb	Black Fast	11	5:01	
7-Feb		12	4:57	
8-Feb		13	4:53	
9-Feb	Full Moon	14	4:49	Appulse Lunar Eclipse 2:39 PM
10-Feb		15	4:45	
11-Feb		16	4:42	
12-Feb		17	4:38	
13-Feb		18	4:34	
14-Feb		19	4:30	
15-Feb		20	4:26	
16-Feb		21	4:22	
17-Feb		22	4:18	
18-Feb		23	4:14	
19-Feb		24	4:10	
20-Feb		25	4:06	
21-Feb		26	4:02	
22-Feb		27	3:58	
23-Feb		28	3:54	
24-Feb		29	3:50	
25-Feb	New Moon	0	3:46	
26-Feb		1	3:42	

Date	Day of Walking	Day of the Moon	Time of Calling	Notes (*Times in Ephemeris Time*)
27-Feb		2	3:38 AM	
28-Feb		3	3:34	
1-Mar		4	3:30	
2-Mar		5	3:26	
3-Mar		6	3:22	
4-Mar		7	3:18	
5-Mar	Fast Begins	8	3:14	
6-Mar		9	3:10	
7-Mar		10	3:06	
8-Mar	Black Fast	11	3:02	
9-Mar		12	2:58	
10-Mar		13	2:54	
11-Mar	Full Moon	14	2:50	
12-Mar		15	2:46	
13-Mar		16	2:42	
14-Mar		17	2:38	
15-Mar		18	2:34	
16-Mar		19	2:30	
17-Mar		20	2:26	
18-Mar		21	2:22	
19-Mar		22	2:18	

20-Mar		23	2:14	Vernal Equinox 11:45 AM
21-Mar		24	2:10	
22-Mar		25	2:06	
23-Mar		26	2:02	
24-Mar		27	1:58	
25-Mar		28	1:54	
26-Mar	New Moon	0	1:50	
27-Mar		1	1:46	
28-Mar		2	1:42	
29-Mar		3	1:38	
30-Mar		4	1:34	
31-Mar		5	1:30	
1-Apr		6	1:26	
2-Apr		7	1:22	
3-Apr	Fast Begins	8	1:18	
4-Apr		9	1:14	
5-Apr		10	1:10	
6-Apr	Black Fast	11	1:06	
7-Apr		12	1:02	
8-Apr		13	12:58	
9-Apr	Full Moon	14	12:54	
10-Apr		15	12:50	
11-Apr		16	12:46	
12-Apr		17	12:42	

Date	Day of Walking	Day of the Moon	Time of Calling	Notes (*Times in Ephemeris Time*)
13-Apr		18	12:38 AM	
14-Apr		19	12:34	
15-Apr		20	12:30	
16-Apr		21	12:25	
17-Apr		22	12:21	
18-Apr		23	12:17	
19-Apr		24	12:13	
20-Apr		25	12:09	
21-Apr		26	12:05	
22-Apr		27	12:01 AM	
23-Apr		28	11:57 PM	
24-Apr		29	11:53	
25-Apr	New Moon	0	11:49	
26-Apr		1	11:45	
27-Apr		2	11:41	
28-Apr		3	11:37	
29-Apr		4	11:33	
30-Apr		5	11:30	Beltane
1-May		6	11:26	
2-May		7	11:22	
3-May	Fast Begins	8	11:18	

4-May		9	11:14
5-May		10	11:10
6-May	Black Fast	11	11:06
7-May		12	11:02
8-May		13	10:58
9-May	Full Moon	14	10:54
10-May		15	10:50
11-May		16	10:46
12-May		17	10:42
13-May		18	10:38
14-May		19	10:34
15-May		20	10:31
16-May		21	10:27
17-May		22	10:23
18-May		23	10:19
19-May		24	10:15
20-May		25	10:11
21-May		26	10:07
22-May		27	10:03
23-May		28	9:59
24-May	New Moon	0	9:55
25-May		1	9:51
26-May		2	9:47
27-May		3	9:43

Date	Day of Walking	Day of the Moon	Time of Calling	Notes (*Times in Ephemeris Time*)
28-May		4	9:39 PM	
29-May		5	9:35	
30-May		6	9:31	
31-May		7	9:28	
1-Jun	Fast Begins	8	9:24	
2-Jun		9	9:20	
3-Jun		10	9:16	
4-Jun	Black Fast	11	9:12	
5-Jun		12	9:08	
6-Jun		13	9:04	
7-Jun	Full Moon	14	9:00	
8-Jun		15	8:56	
9-Jun		16	8:52	
10-Jun		17	8:48	
11-Jun		18	8:44	
12-Jun		19	8:40	
13-Jun		20	8:36	
14-Jun		21	8:32	
15-Jun		22	8:28	
16-Jun		23	8:25	
17-Jun		24	8:21	

18-Jun		25	8:17	
19-Jun		26	8:13	
20-Jun		27	8:09	
21-Jun		28	8:05	Summer Solstice 5:47 AM
22-Jun	New Moon	0	8:01	
23-Jun		1	7:57	
24-Jun		2	7:53	
25-Jun		3	7:49	
26-Jun		4	7:45	
27-Jun		5	7:41	
28-Jun		6	7:37	
29-Jun		7	7:33	
30-Jun		8	7:29	
1-Jul	Fast Begins	9	7:25	
2-Jul		10	7:22	
3-Jul		11	7:18	
4-Jul	Black Fast	12	7:14	
5-Jul		13	7:10	
6-Jul		14	7:06	
7-Jul	Full Moon	15	7:02	Appulse Lunar Eclipse 9:40 AM
8-Jul		16	6:58	
9-Jul		17	6:54	
10-Jul		18	6:50	
11-Jul		19	6:46	

Date	Day of Walking	Day of the Moon	Time of Calling	Notes *(Times in Ephemeris Time)*
12-Jul		20	6:42 PM	
13-Jul		21	6:38	
14-Jul		22	6:34	
15-Jul		23	6:30	
16-Jul		24	6:26	
17-Jul		25	6:22	
18-Jul		26	6:19	
19-Jul		27	6:15	
20-Jul		28	6:11	
21-Jul		29	6:07	
22-Jul	New Moon	0	6:03	Total Solar Eclipse 2:36 AM
23-Jul		1	5:59	
24-Jul		2	5:55	
25-Jul		3	5:51	
26-Jul		4	5:47	
27-Jul		5	5:43	
28-Jul		6	5:39	
29-Jul		7	5:35	
30-Jul		8	5:31	
31-Jul	Fast Begins	9	5:27	
1-Aug		10	5:23	Lammas

2-Aug		11	5:19	
3-Aug	Black Fast	12	5:18	
4-Aug		13	5:14	
5-Aug		14	5:10	
6-Aug	Full Moon	15	5:06	Appulse Lunar Eclipse 12:40 AM
7-Aug		16	5:02	
8-Aug		17	4:58	
9-Aug		18	4:54	
10-Aug		19	4:50	
11-Aug		20	4:46	
12-Aug		21	4:42	
13-Aug		22	4:38	
14-Aug		23	4:34	
15-Aug		24	4:30	
16-Aug		25	4:26	
17-Aug		26	4:22	
18-Aug		27	4:18	
19-Aug		28	4:16	
20-Aug	New Moon	0	4:12	
21-Aug		1	4:08	
22-Aug		2	4:04	
23-Aug		3	4:00	
24-Aug		4	3:56	
25-Aug		5	3:52	

Date	Day of Walking	Day of the Moon	Time of Calling	Notes (*Times in Ephemeris Time*)
26-Aug		6	3:48 PM	
27-Aug		7	3:44	
28-Aug		8	3:40	
29-Aug	Fast Begins	9	3:36	
30-Aug		10	3:32	
31-Aug		11	3:28	
1-Sep	Black Fast	12	3:24	
2-Sep		13	3:20	
3-Sep		14	3:16	
4-Sep	Full Moon	15	3:12	
5-Sep		16	3:09	
6-Sep		17	3:05	
7-Sep		18	3:01	
8-Sep		19	2:57	
9-Sep		20	2:53	
10-Sep		21	2:49	
11-Sep		22	2:45	
12-Sep		23	2:41	
13-Sep		24	2:37	
14-Sep		25	2:33	
15-Sep		26	2:29	

16-Sep		27	2:25	
17-Sep		28	2:21	
18-Sep	New Moon	0	2:17	
19-Sep		1	2:13	
20-Sep		2	2:09	
21-Sep		3	2:06	
22-Sep		4	2:02	Autumnal Equinox 9:20 PM
23-Sep		5	1:58	
24-Sep		6	1:54	
25-Sep		7	1:50	
26-Sep		8	1:46	
27-Sep		9	1:42	
28-Sep	Fast Begins	10	1:38	
29-Sep		11	1:34	
30-Sep		12	1:30	
1-Oct	Black Fast	13	1:26	
2-Oct		14	1:22	
3-Oct		15	1:18	
4-Oct	Full Moon	16	1:14	
5-Oct		17	1:10	
6-Oct		18	1:06	
7-Oct		19	1:03	
8-Oct		20	12:59	
9-Oct		21	12:55	

Date	Day of Walking	Day of the Moon	Time of Calling	Notes *(Times in Ephemeris Time)*
10-Oct		22	12:51 PM	
11-Oct		23	12:47	
12-Oct		24	12:43	
13-Oct		25	12:39	
14-Oct		26	12:35	
15-Oct		27	12:31	
16-Oct		28	12:27	
17-Oct		29	12:23	
18-Oct	New Moon	0	12:19	
19-Oct		1	12:15	
20-Oct		2	12:11	
21-Oct		3	12:07	
22-Oct		4	12:03	
23-Oct		5	12:00 PM	
24-Oct		6	11:56 AM	
25-Oct		7	11:52	
26-Oct		8	11:48	
27-Oct	Fast Begins	9	11:44	
28-Oct		10	11:40	
29-Oct		11	11:36	
30-Oct	Black Fast	12	11:32	

31-Oct		13	11:28	Samhain
1-Nov		14	11:24	
2-Nov	Full Moon	15	11:20	
3-Nov		16	11:16	
4-Nov		17	11:12	
5-Nov		18	11:08	
6-Nov		19	11:04	
7-Nov		20	11:00	
8-Nov		21	10:57	
9-Nov		22	10:53	
10-Nov		23	10:49	
11-Nov		24	10:45	
12-Nov		25	10:41	
13-Nov		26	10:37	
14-Nov		27	10:33	
15-Nov		28	10:29	
16-Nov	New Moon	0	10:25	
17-Nov		1	10:21	
18-Nov		2	10:17	
19-Nov		3	10:13	
20-Nov		4	10:09	
21-Nov		5	10:05	
22-Nov		6	10:01	
23-Nov		7	9:57	

Date	Day of Walking	Day of the Moon	Time of Calling	Notes *(Times in Ephemeris Time)*
24-Nov		8	9:54 AM	
25-Nov		9	9:50	
26-Nov	Fast Begins	10	9:46	
27-Nov		11	9:42	
28-Nov		12	9:38	
29-Nov	Black Fast	13	9:34	
30-Nov		14	9:30	
1-Dec		15	9:26	
2-Dec	Full Moon	16	9:22	
3-Dec		17	9:18	
4-Dec		18	9:14	
5-Dec		19	9:10	
6-Dec		20	9:06	
7-Dec		21	9:02	
8-Dec		22	8:58	
9-Dec		23	8:54	
10-Dec		24	8:50	
11-Dec		25	8:46	
12-Dec		26	8:42	
13-Dec		27	8:38	
14-Dec		28	8:34	

	15-Dec		29	8:30	
	16-Dec	New Moon	0	8:26	
	17-Dec		1	8:22	
	18-Dec		2	8:18	
	19-Dec		3	8:14	
	20-Dec		4	8:10	
	21-Dec		5	8:06	Winter Solstice 5:48 PM
	22-Dec		6	8:02	
	23-Dec		7	7:58	
	24-Dec		8	7:54	
	25-Dec	Fast Begins	9	7:51	
	26-Dec		10	7:47	
	27-Dec		11	7:43	
	28-Dec	Black Fast	12	7:39	
	29-Dec		13	7:35	
	30-Dec		14	7:31	
	31-Dec	Full Moon	15	7:27	Partial Lunar Eclipse 7:24 PM
2010	1-Jan		16	7:23	
	2-Jan		17	7:19	
	3-Jan		18	7:15	
	4-Jan		19	7:11	
	5-Jan		20	7:07	
	6-Jan		21	7:03	

Date	Day of Walking	Day of the Moon	Time of Calling	Notes *(Times in Ephemeris Time)*
7-Jan		22	6:59 AM	
8-Jan		23	6:55	
9-Jan		24	6:51	
10-Jan		25	6:47	
11-Jan		26	6:44	
12-Jan		27	6:40	
13-Jan		28	6:36	
14-Jan		29	6:32	
15-Jan	New Moon	0	6:28	Annular Solar Eclipse 7:07 AM
16-Jan		1	6:24	
17-Jan		2	6:20	
18-Jan		3	6:16	
19-Jan		4	6:12	
20-Jan		5	6:08	
21-Jan		6	6:04	
22-Jan		7	6:00	
23-Jan		8	5:56	
24-Jan	Fast Begins	9	5:52	
25-Jan		10	5:48	
26-Jan		11	5:45	
27-Jan	Black Fast	12	5:41	

28-Jan		13	5:37	
29-Jan		14	5:33	
30-Jan	Full Moon	15	5:29	
31-Jan		16	5:25	Imbolc
1-Feb		17	5:21	
2-Feb		18	5:17	
3-Feb		19	5:13	
4-Feb		20	5:09	
5-Feb		21	5:05	
6-Feb		22	5:01	
7-Feb		23	4:57	
8-Feb		24	4:53	
9-Feb		25	4:49	
10-Feb		26	4:45	
11-Feb		27	4:42	
12-Feb		28	4:38	
13-Feb		29	4:34	
14-Feb	New Moon	0	4:30	
15-Feb		1	4:26	
16-Feb		2	4:22	
17-Feb		3	4:18	
18-Feb		4	4:14	
19-Feb		5	4:10	
20-Feb		6	4:06	

Date	Day of Walking	Day of the Moon	Time of Calling	Notes *(Times in Ephemeris Time)*
21-Feb		7	4:02 AM	
22-Feb	Fast Begins	8	3:58	
23-Feb		9	3:54	
24-Feb		10	3:50	
25-Feb	Black Fast	11	3:46	
26-Feb		12	3:42	
27-Feb		13	3:38	
28-Feb	Full Moon	14	3:34	
1-Mar		15	3:30	
2-Mar		16	3:26	
3-Mar		17	3:22	
4-Mar		18	3:18	
5-Mar		19	3:14	
6-Mar		20	3:10	
7-Mar		21	3:06	
8-Mar		22	3:02	
9-Mar		23	2:58	
10-Mar		24	2:54	
11-Mar		25	2:50	
12-Mar		26	2:46	
13-Mar		27	2:42	

14-Mar		28	2:38	
15-Mar	New Moon	0	2:34	
16-Mar		1	2:30	
17-Mar		2	2:26	
18-Mar		3	2:22	
19-Mar		4	2:18	
20-Mar		5	2:14	Vernal Equinox 5:33 PM
21-Mar		6	2:10	
22-Mar		7	2:06	
23-Mar		8	2:02	
24-Mar	Fast Begins	9	1:58	
25-Mar		10	1:54	
26-Mar		11	1:50	
27-Mar	Black Fast	12	1:46	
28-Mar		13	1:42	
29-Mar		14	1:38	
30-Mar	Full Moon	15	1:34	
31-Mar		16	1:30	
1-Apr		17	1:26	
2-Apr		18	1:22	
3-Apr		19	1:18	
4-Apr		20	1:14	
5-Apr		21	1:10	
6-Apr		22	1:06	

Date	Day of Walking	Day of the Moon	Time of Calling	Notes (*Times in Ephemeris Time*)
7-Apr		23	1:02 AM	
8-Apr		24	12:58	
9-Apr		25	12:54	
10-Apr		26	12:50	
11-Apr		27	12:46	
12-Apr		28	12:42	
13-Apr		29	12:38	
14-Apr	New Moon	0	12:34	
15-Apr		1	12:30	
16-Apr		2	12:25	
17-Apr		3	12:21	
18-Apr		4	12:17	
19-Apr		5	12:13	
20-Apr		6	12:09	
21-Apr		7	12:05	
22-Apr	Fast Begins	8	12:01 AM	
23-Apr		9	11:57 PM	
24-Apr		10	11:53	
25-Apr	Black Fast	11	11:49	
26-Apr		12	11:45	
27-Apr		13	11:41	

28-Apr	Full Moon	14	11:37	
29-Apr		15	11:33	
30-Apr		16	11:30	Beltane
1-May		17	11:26	
2-May		18	11:22	
3-May		19	11:18	
4-May		20	11:14	
5-May		21	11:10	
6-May		22	11:06	
7-May		23	11:02	
8-May		24	10:58	
9-May		25	10:54	
10-May		26	10:50	
11-May		27	10:46	
12-May		28	10:42	
13-May		29	10:38	
14-May	New Moon	0	10:34	
15-May		1	10:31	
16-May		2	10:27	
17-May		3	10:23	
18-May		4	10:19	
19-May		5	10:15	
20-May		6	10:11	
21-May	Fast Begins	7	10:07	

Date	Day of Walking	Day of the Moon	Time of Calling	Notes *(Times in Ephemeris Time)*
22-May		8	10:03 PM	
23-May		9	9:59	
24-May	Black Fast	10	9:55	
25-May		11	9:51	
26-May		12	9:47	
27-May	Full Moon	13	9:43	
28-May		14	9:39	
29-May		15	9:35	
30-May		16	9:31	
31-May		17	9:28	
1-Jun		18	9:24	
2-Jun		19	9:20	
3-Jun		20	9:16	
4-Jun		21	9:12	
5-Jun		22	9:08	
6-Jun		23	9:04	
7-Jun		24	9:00	
8-Jun		25	8:56	
9-Jun		26	8:52	
10-Jun		27	8:48	
11-Jun		28	8:44	

12-Jun	New Moon	0	8:40	
13-Jun		1	8:36	
14-Jun		2	8:32	
15-Jun		3	8:28	
16-Jun		4	8:25	
17-Jun		5	8:21	
18-Jun		6	8:17	
19-Jun		7	8:13	
20-Jun	Fast Begins	8	8:09	
21-Jun		9	8:05	Summer Solstice 11:30 AM
22-Jun		10	8:01	
23-Jun	Black Fast	11	7:57	
24-Jun		12	7:53	
25-Jun		13	7:49	
26-Jun	Full Moon	14	7:45	Partial Lunar Eclipse 11:40 AM
27-Jun		15	7:41	
28-Jun		16	7:37	
29-Jun		17	7:33	
30-Jun		18	7:29	
1-Jul		19	7:25	
2-Jul		20	7:22	
3-Jul		21	7:18	
4-Jul		22	7:14	
5-Jul		23	7:10	

Date	Day of Walking	Day of the Moon	Time of Calling	Notes *(Times in Ephemeris Time)*
6-Jul		24	7:06 PM	
7-Jul		25	7:02	
8-Jul		26	6:58	
9-Jul		27	6:54	
10-Jul		28	6:50	
11-Jul	New Moon	0	6:46	Total Solar Eclipse 7:34 PM
12-Jul		1	6:42	
13-Jul		2	6:38	
14-Jul		3	6:34	
15-Jul		4	6:30	
16-Jul		5	6:26	
17-Jul		6	6:22	
18-Jul		7	6:19	
19-Jul		8	6:15	
20-Jul	Fast Begins	9	6:11	
21-Jul		10	6:07	
22-Jul		11	6:03	
23-Jul	Black Fast	12	5:59	
24-Jul		13	5:55	
25-Jul		14	5:51	
26-Jul	Full Moon	15	5:47	

27-Jul		16	5:43	
28-Jul		17	5:39	
29-Jul		18	5:35	
30-Jul		19	5:31	
31-Jul		20	5:27	
1-Aug		21	5:23	Lammas
2-Aug		22	5:19	
3-Aug		23	5:18	
4-Aug		24	5:14	
5-Aug		25	5:10	
6-Aug		26	5:06	
7-Aug		27	5:02	
8-Aug		28	4:58	
9-Aug		29	4:54	
10-Aug	New Moon	0	4:50	
11-Aug		1	4:46	
12-Aug		2	4:42	
13-Aug		3	4:38	
14-Aug		4	4:34	
15-Aug		5	4:30	
16-Aug		6	4:26	
17-Aug		7	4:22	
18-Aug	Fast Begins	8	4:18	
19-Aug		9	4:16	

Date	Day of Walking	Day of the Moon	Time of Calling	Notes (Times in Ephemeris Time)
20-Aug		10	4:12 PM	
21-Aug	Black Fast	11	4:08	
22-Aug		12	4:04	
23-Aug		13	4:00	
24-Aug	Full Moon	14	3:56	
25-Aug		15	3:52	
26-Aug		16	3:48	
27-Aug		17	3:44	
28-Aug		18	3:40	
29-Aug		19	3:36	
30-Aug		20	3:32	
31-Aug		21	3:28	
1-Sep		22	3:24	
2-Sep		23	3:20	
3-Sep		24	3:16	
4-Sep		25	3:12	
5-Sep		26	3:09	
6-Sep		27	3:05	
7-Sep		28	3:01	
8-Sep	New Moon	0	2:57	
9-Sep		1	2:53	

10-Sep		2	2:49	
11-Sep		3	2:45	
12-Sep		4	2:41	
13-Sep		5	2:37	
14-Sep		6	2:33	
15-Sep		7	2:29	
16-Sep		8	2:25	
17-Sep	Fast Begins	9	2:21	
18-Sep		10	2:17	
19-Sep		11	2:13	
20-Sep	Black Fast	12	2:09	
21-Sep		13	2:06	
22-Sep		14	2:02	
23-Sep	Full Moon	15	1:58	Autumnal Equinox 3:10 AM
24-Sep		16	1:54	
25-Sep		17	1:50	
26-Sep		18	1:46	
27-Sep		19	1:42	
28-Sep		20	1:38	
29-Sep		21	1:34	
30-Sep		22	1:30	
1-Oct		23	1:26	
2-Oct		24	1:22	
3-Oct		25	1:18	

Date	Day of Walking	Day of the Moon	Time of Calling	Notes (*Times in Ephemeris Time*)
4-Oct		26	1:14 PM	
5-Oct		27	1:10	
6-Oct		28	1:06	
7-Oct	New Moon	0	1:03	
8-Oct		1	12:59	
9-Oct		2	12:55	
10-Oct		3	12:51	
11-Oct		4	12:47	
12-Oct		5	12:43	
13-Oct		6	12:39	
14-Oct		7	12:35	
15-Oct		8	12:31	
16-Oct		9	12:27	
17-Oct	Fast Begins	10	12:23	
18-Oct		11	12:19	
19-Oct		12	12:15	
20-Oct	Black Fast	13	12:11	
21-Oct		14	12:07	
22-Oct		15	12:03	
23-Oct	Full Moon	16	12:00 PM	
24-Oct		17	11:56 AM	

25-Oct		18	11:52	
26-Oct		19	11:48	
27-Oct		20	11:44	
28-Oct		21	11:40	
29-Oct		22	11:36	
30-Oct		23	11:32	
31-Oct		24	11:28	Samhain
1-Nov		25	11:24	
2-Nov		26	11:20	
3-Nov		27	11:16	
4-Nov		28	11:12	
5-Nov		29	11:08	
6-Nov	New Moon	0	11:04	
7-Nov		1	11:00	
8-Nov		2	10:57	
9-Nov		3	10:53	
10-Nov		4	10:49	
11-Nov		5	10:45	
12-Nov		6	10:41	
13-Nov		7	10:37	
14-Nov		8	10:33	
15-Nov	Fast Begins	9	10:29	
16-Nov		10	10:25	
17-Nov		11	10:21	

Date	Day of Walking	Day of the Moon	Time of Calling	Notes (Times in Ephemeris Time)
18-Nov	Black Fast	12	10:17 AM	
19-Nov		13	10:13	
20-Nov		14	10:09	
21-Nov	Full Moon	15	10:05	
22-Nov		16	10:01	
23-Nov		17	9:57	
24-Nov		18	9:54	
25-Nov		19	9:50	
26-Nov		20	9:46	
27-Nov		21	9:42	
28-Nov		22	9:38	
29-Nov		23	9:34	
30-Nov		24	9:30	
1-Dec		25	9:26	
2-Dec		26	9:22	
3-Dec		27	9:18	
4-Dec		28	9:14	
5-Dec	New Moon	0	9:10	
6-Dec		1	9:06	
7-Dec		2	9:02	
8-Dec		3	8:58	

9-Dec		4	8:54	
10-Dec		5	8:50	
11-Dec		6	8:46	
12-Dec		7	8:42	
13-Dec		8	8:38	
14-Dec		9	8:34	
15-Dec	Fast Begins	10	8:30	
16-Dec		11	8:26	
17-Dec		12	8:22	
18-Dec	Black Fast	13	8:18	
19-Dec		14	8:14	
20-Dec		15	8:10	
21-Dec	Full Moon	16	8:06	Winter Solstice 11:40 PM/ Total Lunar Eclipse 8:18 AM
22-Dec		17	8:02	
23-Dec		18	7:58	
24-Dec		19	7:54	
25-Dec		20	7:51	
26-Dec		21	7:47	
27-Dec		22	7:43	
28-Dec		23	7:39	
29-Dec		24	7:35	
30-Dec		25	7:31	
31-Dec		26	7:27	

Date		Day of Walking	Day of the Moon	Time of Calling	Notes *(Times in Ephemeris Time)*
2011	1-Jan		27	7:23 AM	
	2-Jan		28	7:19	
	3-Jan		29	7:15	
	4-Jan	New Moon	0	7:11	Partial Solar Eclipse 8:51 AM
	5-Jan		1	7:07	
	6-Jan		2	7:03	
	7-Jan		3	6:59	
	8-Jan		4	6:55	
	9-Jan		5	6:51	
	10-Jan		6	6:47	
	11-Jan		7	6:44	
	12-Jan		8	6:40	
	13-Jan	Fast Begins	9	6:36	
	14-Jan		10	6:32	
	15-Jan		11	6:28	
	16-Jan	Black Fast	12	6:24	
	17-Jan		13	6:20	
	18-Jan		14	6:16	
	19-Jan	Full Moon	15	6:12	
	20-Jan		16	6:08	
	21-Jan		17	6:04	

22-Jan		18	6:00	
23-Jan		19	5:56	
24-Jan		20	5:52	
25-Jan		21	5:48	
26-Jan		22	5:45	
27-Jan		23	5:41	
28-Jan		24	5:37	
29-Jan		25	5:33	
30-Jan		26	5:29	
31-Jan		27	5:25	Imbolc
1-Feb		28	5:21	
2-Feb		29	5:17	
3-Feb	New Moon	0	5:13	
4-Feb		1	5:09	
5-Feb		2	5:05	
6-Feb		3	5:01	
7-Feb		4	4:57	
8-Feb		5	4:53	
9-Feb		6	4:49	
10-Feb		7	4:45	
11-Feb		8	4:42	
12-Feb	Fast Begins	9	4:38	
13-Feb		10	4:34	
14-Feb		11	4:30	

Date	Day of Walking	Day of the Moon	Time of Calling	Notes (*Times in Ephemeris Time*)
15-Feb	Black Fast	12	4:26 AM	
16-Feb		13	4:22	
17-Feb		14	4:18	
18-Feb	Full Moon	15	4:14	
19-Feb		16	4:10	
20-Feb		17	4:06	
21-Feb		18	4:02	
22-Feb		19	3:58	
23-Feb		20	3:54	
24-Feb		21	3:50	
25-Feb		22	3:46	
26-Feb		23	3:42	
27-Feb		24	3:38	
28-Feb		25	3:34	
1-Mar		26	3:30	
2-Mar		27	3:26	
3-Mar		28	3:22	
4-Mar	New Moon	0	3:18	
5-Mar		1	3:14	
6-Mar		2	3:10	
7-Mar		3	3:06	

8-Mar		4	3:02	
9-Mar		5	2:58	
10-Mar		6	2:54	
11-Mar		7	2:50	
12-Mar		8	2:46	
13-Mar	Fast Begins	9	2:42	
14-Mar		10	2:38	
15-Mar		11	2:34	
16-Mar	Black Fast	12	2:30	
17-Mar		13	2:26	
18-Mar		14	2:22	
19-Mar	Full Moon	15	2:18	
20-Mar		16	2:14	Vernal Equinox 11:22 PM
21-Mar		17	2:10	
22-Mar		18	2:06	
23-Mar		19	2:02	
24-Mar		20	1:58	
25-Mar		21	1:54	
26-Mar		22	1:50	
27-Mar		23	1:46	
28-Mar		24	1:42	
29-Mar		25	1:38	
30-Mar		26	1:34	
31-Mar		27	1:30	

Date	Day of Walking	Day of the Moon	Time of Calling	Notes *(Times in Ephemeris Time)*
1-Apr		28	1:26 AM	
2-Apr		29	1:22	
3-Apr	New Moon	0	1:18	
4-Apr		1	1:14	
5-Apr		2	1:10	
6-Apr		3	1:06	
7-Apr		4	1:02	
8-Apr		5	12:58	
9-Apr		6	12:54	
10-Apr		7	12:50	
11-Apr		8	12:46	
12-Apr	Fast Begins	9	12:42	
13-Apr		10	12:38	
14-Apr		11	12:34	
15-Apr	Black Fast	12	12:30	
16-Apr		13	12:25	
17-Apr		14	12:21	
18-Apr	Full Moon	15	12:17	
19-Apr		16	12:13	
20-Apr		17	12:09	
21-Apr		18	12:05	

22-Apr		19	12:01 AM	
23-Apr		20	11:57 PM	
24-Apr		21	11:53	
25-Apr		22	11:49	
26-Apr		23	11:45	
27-Apr		24	11:41	
28-Apr		25	11:37	
29-Apr		26	11:33	
30-Apr		27	11:30	Beltane
1-May		28	11:26	
2-May		29	11:22	
3-May	New Moon	0	11:18	
4-May		1	11:14	
5-May		2	11:10	
6-May		3	11:06	
7-May		4	11:02	
8-May		5	10:58	
9-May		6	10:54	
10-May		7	10:50	
11-May	Fast Begins	8	10:46	
12-May		9	10:42	
13-May		10	10:38	
14-May	Black Fast	11	10:34	
15-May		12	10:31	

Date	Day of Walking	Day of the Moon	Time of Calling	Notes (*Times in Ephemeris Time*)
16-May		13	10:27 PM	
17-May	Full Moon	14	10:23	
18-May		15	10:19	
19-May		16	10:15	
20-May		17	10:11	
21-May		18	10:07	
22-May		19	10:03	
23-May		20	9:59	
24-May		21	9:55	
25-May		22	9:51	
26-May		23	9:47	
27-May		24	9:43	
28-May		25	9:39	
29-May		26	9:35	
30-May		27	9:31	
31-May		28	9:28	
1-Jun	New Moon	0	9:24	Partial Solar Eclipse 9:17 PM
2-Jun		1	9:20	
3-Jun		2	9:16	
4-Jun		3	9:12	
5-Jun		4	9:08	

Davies, Paul, *The Mind of God: The Scientific Basis for a Rational World*, Simon & Schuster, NY, 1992

Eliade, Mircea, *Rites and Symbols of Initiation*, Harper Torchbooks, NY, 1965

Evans-Wentz, W. Y., *The Tibetan Book of the Dead*, Oxford University Press, NY, 1969

Horgan, John, "Profile: Francis H.C. Crick, The Mephistopheles of Neurobiology," in *Scientific American*, February 1992, vol. 266, No. 2

Jung, C. G. (ed.) *Man and His Symbols*, Doubleday, NY, 1964

— *Essays on a Science of Mythology*, Bollingen Series, Princeton, 1989

Kerenyi, C., and Jung, C. G., *Essays on a Science of Mythology*, Bollingen Princeton, 1989

Kohn, Livia, ed., *Taoist Meditation and Longevity Techniques,* The University of Michigan, Ann Arbor, 1989

Kramer, S. N., *The Sumerians*, University of Chicago Press, Chicago, 1963

Krupp, E. C., *Echoes of the Ancient Skies*, Harper & Row, NY, 1983

Laing, R. D., *The Politics of Experience*, Pantheon Books, NY, 1967

Lamy, Lucie, *Egyptian Mysteries*, Thames & Hudson, NY, 1981

MacCana, Proinsias, *Celtic Mythology*, Peter Bedrick Books, NY, 1985

Neely, Henry M., *A Primer for Star-Gazers*, Harper & Row, NY, 1989

Parkinson, C. N., *Parkinson's Law and Other Studies in Administration,* Ballantine Books, NY, 1957

Robinet, Isabelle, "Visualization and Ecstatic Flight in Shangqing Daoism," in Kohn, L. (ed.), op. cit.

Strickmann, Michel, "On the Alchemy of T'ao Hung-Ching," in Welch & Seidel (1979)

Taylor, Timothy, "The Gundestrup Cauldron" in *Scientific American*, March 1992, vol. 266, No. 3, 84–89

Temple, R.K.G., *The Sirius Mystery*, St. Martin's Press, NY, 1976

Von Franz, M. L., "The Process of Individuation" in Jung, C.G. (1964), 171

Walters, Derek, *Chinese Astrology*, The Aquarian Press, 1987

Watkins, C., (ed.), *The American Heritage Dictionary of Indo- European Roots*, Houghton Mifflin Co., Boston, 1985

Welch, Holmes, and Seidel, Anna, (eds.), *Facets of Taoism*, Yale University Press, NY, 1979

Recommended Reading

Campbell, J. (ed.), *Spiritual Disciplines: Papers from the Eranos Yearbooks*, Bollingen Series, Princeton, 1985

Dimock, E. C., Jr., *The Place of the Hidden Moon*, University of Chicago Press, Chicago, 1989

Eliade, Mircea, *The Quest: History and Meaning in Religion*, University of Chicago Press, Chicago, 1975

Frazer, James G., *The Golden Bough*, Macmillan Co., NY, 1922

Gay, Peter, *Freud: A Life For Our Time*, Anchor Books, NY, 1988

Krupp, E. C. (ed.), *In Search of Ancient Astronomies*, Doubleday & Co., NY, 1977

Lindsay, Jack, *The Origins of Alchemy in Graeco-Roman Egypt*, Barnes & Noble, NY, 1970

Luk, Charles, *Taoist Yoga*, Samuel Weiser, York Beach, Maine, 1984

Maspero, Gaston, *The Dawn of Civilization*, Society for Promoting Christian Knowledge, London, 1922

Neumann, Erich, *The Great Mother*, Bollingen Series, Princeton, 1974

Nicholson, Irene, *Mexican And Central American Mythology*, Peter Bedrick Books, NY, 1985

Regardie, Israel, *The Golden Dawn*, Llewellyn Publications, St. Paul, 1986

Skinner, Stephen, *The Living Earth Manual of Feng-Shui*, Routledge & Kegan Paul, London, 1982

Turner, Robert, *The Heptarchia Mystica of John Dee,* Aquarian Press, Wellingborough, 1986

— *Elizabethan Magic*, Element Books, Dorset, 1989

Walker, Benjamin, *Gnosticism: Its History and Influence*, Aquarian Press, Wellingborough, 1983

Welch, Holmes, *Taoism: The Parting of the Way*, Beacon Press, Boston, 1966

Zimmer, Heinrich, *Myths and Symbols in Indian Art and Civilization*, Bollingen Series, Princeton, 1974

6-Jun		5	9:04	
7-Jun		6	9:00	
8-Jun		7	8:56	
9-Jun	Fast Begins	8	8:52	
10-Jun		9	8:48	
11-Jun		10	8:44	
12-Jun	Black Fast	11	8:40	
13-Jun		12	8:36	
14-Jun		13	8:32	
15-Jun	Full Moon	14	8:28	Total Lunar Eclipse 8:14 PM
16-Jun		15	8:25	
17-Jun		16	8:21	
18-Jun		17	8:17	
19-Jun		18	8:13	
20-Jun		19	8:09	
21-Jun		20	8:05	Summer Solstice 5:18 PM
22-Jun		21	8:01	
23-Jun		22	7:57	
24-Jun		23	7:53	
25-Jun		24	7:49	
26-Jun		25	7:45	
27-Jun		26	7:41	
28-Jun		27	7:37	
29-Jun		28	7:33	

Date	Day of Walking	Day of the Moon	Time of Calling	Notes *(Times in Ephemeris Time)*
30-Jun		29	7:29 PM	
1-Jul	New Moon	0	7:25	Partial Solar Eclipse 8:39 AM
2-Jul		1	7:22	
3-Jul		2	7:18	
4-Jul		3	7:14	
5-Jul		4	7:10	
6-Jul		5	7:06	
7-Jul		6	7:02	
8-Jul		7	6:58	
9-Jul	Fast Begins	8	6:54	
10-Jul		9	6:50	
11-Jul		10	6:46	
12-Jul	Black Fast	11	6:42	
13-Jul		12	6:38	
14-Jul		13	6:34	
15-Jul	Full Moon	14	6:30	
16-Jul		15	6:26	
17-Jul		16	6:22	
18-Jul		17	6:19	
19-Jul		18	6:15	
20-Jul		19	6:11	

21-Jul		20	6:07	
22-Jul		21	6:03	
23-Jul		22	5:59	
24-Jul		23	5:55	
25-Jul		24	5:51	
26-Jul		25	5:47	
27-Jul		26	5:43	
28-Jul		27	5:39	
29-Jul		28	5:35	
30-Jul	New Moon	0	5:31	
31-Jul		1	5:27	
1-Aug		2	5:23	Lammas
2-Aug		3	5:19	
3-Aug		4	5:18	
4-Aug		5	5:14	
5-Aug		6	5:10	
6-Aug		7	5:06	
7-Aug	Fast Begins	8	5:02	
8-Aug		9	4:58	
9-Aug		10	4:54	
10-Aug	Black Fast	11	4:50	
11-Aug		12	4:46	
12-Aug		13	4:42	
13-Aug	Full Moon	14	4:38	

Date	Day of Walking	Day of the Moon	Time of Calling	Notes (*Times in Ephemeris Time*)
14-Aug		15	4:34 PM	
15-Aug		16	4:30	
16-Aug		17	4:26	
17-Aug		18	4:22	
18-Aug		19	4:18	
19-Aug		20	4:16	
20-Aug		21	4:12	
21-Aug		22	4:08	
22-Aug		23	4:04	
23-Aug		24	4:00	
24-Aug		25	3:56	
25-Aug		26	3:52	
26-Aug		27	3:48	
27-Aug		28	3:44	
28-Aug		29	3:40	
29-Aug	New Moon	0	3:36	
30-Aug		1	3:32	
31-Aug		2	3:28	
1-Sep		3	3:24	
2-Sep		4	3:20	
3-Sep		5	3:16	

4-Sep		6	3:12	
5-Sep		7	3:09	
6-Sep	Fast Begins	8	3:05	
7-Sep		9	3:01	
8-Sep		10	2:57	
9-Sep	Black Fast	11	2:53	
10-Sep		12	2:49	
11-Sep		13	2:45	
12-Sep	Full Moon	14	2:41	
13-Sep		15	2:37	
14-Sep		16	2:33	
15-Sep		17	2:29	
16-Sep		18	2:25	
17-Sep		19	2:21	
18-Sep		20	2:17	
19-Sep		21	2:13	
20-Sep		22	2:09	
21-Sep		23	2:06	
22-Sep		24	2:02	
23-Sep		25	1:58	Autumnal Equinox 9:06 AM
24-Sep		26	1:54	
25-Sep		27	1:50	
26-Sep		28	1:46	
27-Sep	New Moon	0	1:42	

Date	Day of Walking	Day of the Moon	Time of Calling	Notes (*Times in Ephemeris Time*)
28-Sep		1	1:38 PM	
29-Sep		2	1:34	
30-Sep		3	1:30	
1-Oct		4	1:26	
2-Oct		5	1:22	
3-Oct		6	1:18	
4-Oct		7	1:14	
5-Oct		8	1:10	
6-Oct	Fast Begins	9	1:06	
7-Oct		10	1:03	
8-Oct		11	12:59	
9-Oct	Black Fast	12	12:55	
10-Oct		13	12:51	
11-Oct		14	12:47	
12-Oct	Full Moon	15	12:43	
13-Oct		16	12:39	
14-Oct		17	12:35	
15-Oct		18	12:31	
16-Oct		19	12:27	
17-Oct		20	12:23	
18-Oct		21	12:19	

19-Oct		22	12:15	
20-Oct		23	12:11	
21-Oct		24	12:07	
22-Oct		25	12:03	
23-Oct		26	12:00 PM	
24-Oct		27	11:56 AM	
25-Oct		28	11:52	
26-Oct	New Moon	0	11:48	
27-Oct		1	11:44	
28-Oct		2	11:40	
29-Oct		3	11:36	
30-Oct		4	11:32	
31-Oct		5	11:28	Samhain
1-Nov		6	11:24	
2-Nov		7	11:20	
3-Nov		8	11:16	
4-Nov	Fast Begins	9	11:12	
5-Nov		10	11:08	
6-Nov		11	11:04	
7-Nov	Black Fast	12	11:00	
8-Nov		13	10:57	
9-Nov		14	10:53	
10-Nov	Full Moon	15	10:49	
11-Nov		16	10:45	

Date	Day of Walking	Day of the Moon	Time of Calling	Notes (*Times in Ephemeris Time*)
12-Nov		17	10:41 AM	
13-Nov		18	10:37	
14-Nov		19	10:33	
15-Nov		20	10:29	
16-Nov		21	10:25	
17-Nov		22	10:21	
18-Nov		23	10:17	
19-Nov		24	10:13	
20-Nov		25	10:09	
21-Nov		26	10:05	
22-Nov		27	10:01	
23-Nov		28	9:57	
24-Nov		29	9:54	
25-Nov	New Moon	0	9:50	Partial Solar Eclipse 6:21 AM
26-Nov		1	9:46	
27-Nov		2	9:42	
28-Nov		3	9:38	
29-Nov		4	9:34	
30-Nov		5	9:30	
1-Dec		6	9:26	
2-Dec		7	9:22	

3-Dec		8	9:18	
4-Dec	Fast Begins	9	9:14	
5-Dec		10	9:10	
6-Dec		11	9:06	
7-Dec	Black Fast	12	9:02	
8-Dec		13	8:58	
9-Dec		14	8:54	
10-Dec	Full Moon	15	8:50	Total Lunar Eclipse 2:33 PM
11-Dec		16	8:46	
12-Dec		17	8:42	
13-Dec		18	8:38	
14-Dec		19	8:34	
15-Dec		20	8:30	
16-Dec		21	8:26	
17-Dec		22	8:22	
18-Dec		23	8:18	
19-Dec		24	8:14	
20-Dec		25	8:10	
21-Dec		26	8:06	
22-Dec		27	8:02	Winter Solstice 5:31 AM
23-Dec		28	7:58	
24-Dec	New Moon	0	7:54	
25-Dec		1	7:51	
26-Dec		2	7:47	

Date		Day of Walking	Day of the Moon	Time of Calling	Notes (*Times in Ephemeris Time*)
	27-Dec		3	7:43 AM	
	28-Dec		4	7:39	
	29-Dec		5	7:35	
	30-Dec		6	7:31	
	31-Dec		7	7:27	
2012	1-Jan		8	7:23	
	2-Jan		9	7:19	
	3-Jan	Fast Begins	10	7:15	
	4-Jan		11	7:11	
	5-Jan		12	7:07	
	6-Jan	Black Fast	13	7:03	
	7-Jan		14	6:59	
	8-Jan		15	6:55	
	9-Jan	Full Moon	16	6:51	
	10-Jan		17	6:47	
	11-Jan		18	6:44	
	12-Jan		19	6:40	
	13-Jan		20	6:36	
	14-Jan		21	6:32	
	15-Jan		22	6:28	

16-Jan		23	6:24	
17-Jan		24	6:20	
18-Jan		25	6:16	
19-Jan		26	6:12	
20-Jan		27	6:08	
21-Jan		28	6:04	
22-Jan		29	6:00	
23-Jan	New Moon	0	5:56	
24-Jan		1	5:52	
25-Jan		2	5:48	
26-Jan		3	5:45	
27-Jan		4	5:41	
28-Jan		5	5:37	
29-Jan		6	5:33	
30-Jan		7	5:29	
31-Jan		8	5:25	Imbolc
1-Feb	Fast Begins	9	5:21	
2-Feb		10	5:17	
3-Feb		11	5:13	
4-Feb	Black Fast	12	5:09	
5-Feb		13	5:05	
6-Feb		14	5:01	
7-Feb	Full Moon	15	4:57	
8-Feb		16	4:53	

Date	Day of Walking	Day of the Moon	Time of Calling	Notes (*Times in Ephemeris Time*).
9-Feb		17	4:49 AM	
10-Feb		18	4:45	
11-Feb		19	4:42	
12-Feb		20	4:38	
13-Feb		21	4:34	
14-Feb		22	4:30	
15-Feb		23	4:26	
16-Feb		24	4:22	
17-Feb		25	4:18	
18-Feb		26	4:14	
19-Feb		27	4:10	
20-Feb		28	4:06	
21-Feb	New Moon	0	4:02	
22-Feb		1	3:58	
23-Feb		2	3:54	
24-Feb		3	3:50	
25-Feb		4	3:46	
26-Feb		5	3:42	
27-Feb		6	3:38	
28-Feb		7	3:34	
29-Feb		8	3:30	

1-Mar		9	3:26	
2-Mar	Fast Begins	10	3:22	
3-Mar		11	3:18	
4-Mar		12	3:14	
5-Mar	Black Fast	13	3:10	
6-Mar		14	3:06	
7-Mar		15	3:02	
8-Mar	Full Moon	16	2:58	
9-Mar		17	2:54	
10-Mar		18	2:50	
11-Mar		19	2:46	
12-Mar		20	2:42	
13-Mar		21	2:38	
14-Mar		22	2:35	
15-Mar		23	2:32	
16-Mar		24	2:30	
17-Mar		25	2:26	
18-Mar		26	2:22	
19-Mar		27	2:18	
20-Mar		28	2:14	Vernal Equinox 5:16 AM
21-Mar		29	2:10	
22-Mar	New Moon	0	2:06	
23-Mar		1	2:02	
24-Mar		2	1:58	

Date	Day of Walking	Day of the Moon	Time of Calling	Notes (*Times in Ephemeris Time*)
25-Mar		3	1:54 AM	
26-Mar		4	1:50	
27-Mar		5	1:46	
28-Mar		6	1:42	
29-Mar		7	1:38	
30-Mar		8	1:34	
31-Mar	Fast Begins	9	1:30	
1-Apr		10	1:26	
2-Apr		11	1:22	
3-Apr	Black Fast	12	1:18	
4-Apr		13	1:14	
5-Apr		14	1:10	
6-Apr	Full Moon	15	1:06	
7-Apr		16	1:02	
8-Apr		17	12:58	
9-Apr		18	12:54	
10-Apr		19	12:50	
11-Apr		20	12:46	
12-Apr		21	12:42	
13-Apr		22	12:38	
14-Apr		23	12:34	

15-Apr		24	12:30	
16-Apr		25	12:25	
17-Apr		26	12:21	
18-Apr		27	12:17	
19-Apr		28	12:13	
20-Apr		29	12:09	
21-Apr	New Moon	0	12:05	
22-Apr		1	12:01 AM	
23-Apr		2	11:57 PM	
24-Apr		3	11:53	
25-Apr		4	11:49	
26-Apr		5	11:45	
27-Apr		6	11:41	
28-Apr		7	11:37	
29-Apr		8	11:33	
30-Apr	Fast Begins	9	11:30	Beltane
1-May		10	11:26	
2-May		11	11:22	
3-May	Black Fast	12	11:18	
4-May		13	11:14	
5-May		14	11:10	
6-May	Full Moon	15	11:06	
7-May		16	11:02	
8-May		17	10:58	

Date	Day of Walking	Day of the Moon	Time of Calling	Notes (*Times in Ephemeris Time*)
9-May		18	10:54 PM	
10-May		19	10:50	
11-May		20	10:46	
12-May		21	10:42	
13-May		22	10:38	
14-May		23	10:34	
15-May		24	10:31	
16-May		25	10:27	
17-May		26	10:23	
18-May		27	10:19	
19-May		28	10:15	
20-May	New Moon	0	10:11	Annular Solar Eclipse 11:23 PM
21-May		1	10:07	
22-May		2	10:03	
23-May		3	9:59	
24-May		4	9:55	
25-May		5	9:51	
26-May		6	9:47	
27-May		7	9:43	
28-May		8	9:39	
29-May	Fast Begins	9	9:35	

30-May		10	9:31	
31-May		11	9:28	
1-Jun	Black Fast	12	9:24	
2-Jun		13	9:20	
3-Jun		14	9:16	
4-Jun	Full Moon	15	9:12	Partial Lunar Eclipse 11:04 AM
5-Jun		16	9:08	
6-Jun		17	9:04	
7-Jun		18	9:00	
8-Jun		19	8:56	
9-Jun		20	8:52	
10-Jun		21	8:48	
11-Jun		22	8:44	
12-Jun		23	8:40	
13-Jun		24	8:36	
14-Jun		25	8:32	
15-Jun		26	8:28	
16-Jun		27	8:25	
17-Jun		28	8:21	
18-Jun		29	8:17	
19-Jun	New Moon	0	8:13	
20-Jun		1	8:09	Summer Solstice 11:10 PM
21-Jun		2	8:05	
22-Jun		3	8:01	

Date	Day of Walking	Day of the Moon	Time of Calling	Notes (*Times in Ephemeris Time*)
23-Jun		4	7:57 PM	
24-Jun		5	7:53	
25-Jun		6	7:49	
26-Jun		7	7:45	
27-Jun	Fast Begins	8	7:41	
28-Jun		9	7:37	
29-Jun		10	7:33	
30-Jun	Black Fast	11	7:29	
1-Jul		12	7:25	
2-Jul		13	7:22	
3-Jul	Full Moon	14	7:18	
4-Jul		15	7:14	
5-Jul		16	7:10	
6-Jul		17	7:06	
7-Jul		18	7:02	
8-Jul		19	6:58	
9-Jul		20	6:54	
10-Jul		21	6:50	
11-Jul		22	6:46	
12-Jul		23	6:42	
13-Jul		24	6:38	

14-Jul		25	6:34	
15-Jul		26	6:30	
16-Jul		27	6:26	
17-Jul		28	6:22	
18-Jul		29	6:19	
19-Jul	New Moon	0	6:15	
20-Jul		1	6:11	
21-Jul		2	6:07	
22-Jul		3	6:03	
23-Jul		4	5:59	
24-Jul		5	5:55	
25-Jul		6	5:51	
26-Jul		7	5:47	
27-Jul	Fast Begins	8	5:43	
28-Jul		9	5:39	
29-Jul		10	5:35	
30-Jul	Black Fast	11	5:31	
31-Jul		12	5:27	
1-Aug		13	5:23	Lammas
2-Aug	Full Moon	14	5:19	
3-Aug		15	5:18	
4-Aug		16	5:14	
5-Aug		17	5:10	
6-Aug		18	5:06	

Date	Day of Walking	Day of the Moon	Time of Calling	Notes *(Times in Ephemeris Time)*
7-Aug		19	5:02 PM	
8-Aug		20	4:58	
9-Aug		21	4:54	
10-Aug		22	4:50	
11-Aug		23	4:46	
12-Aug		24	4:42	
13-Aug		25	4:38	
14-Aug		26	4:34	
15-Aug		27	4:30	
16-Aug		28	4:26	
17-Aug	New Moon	0	4:22	
18-Aug		1	4:18	
19-Aug		2	4:16	
20-Aug		3	4:12	
21-Aug		4	4:08	
22-Aug		5	4:04	
23-Aug		6	4:00	
24-Aug		7	3:56	
25-Aug	Fast Begins	8	3:52	
26-Aug		9	3:48	
27-Aug		10	3:44	

28-Aug	Black Fast	11	3:40	
29-Aug		12	3:36	
30-Aug		13	3:32	
31-Aug	Full Moon	14	3:28	
1-Sep		15	3:24	
2-Sep		16	3:20	
3-Sep		17	3:16	
4-Sep		18	3:12	
5-Sep		19	3:09	
6-Sep		20	3:05	
7-Sep		21	3:01	
8-Sep		22	2:57	
9-Sep		23	2:53	
10-Sep		24	2:49	
11-Sep		25	2:45	
12-Sep		26	2:41	
13-Sep		27	2:37	
14-Sep		28	2:33	
15-Sep		29	2:29	
16-Sep	New Moon	0	2:25	
17-Sep		1	2:21	
18-Sep		2	2:17	
19-Sep		3	2:13	
20-Sep		4	2:09	

Date	Day of Walking	Day of the Moon	Time of Calling	Notes (*Times in Ephemeris Time*)
21-Sep		5	2:06 PM	
22-Sep		6	2:02	Autumnal Equinox 2:50 PM
23-Sep		7	1:58	
24-Sep	Fast Begins	8	1:54	
25-Sep		9	1:50	
26-Sep		10	1:46	
27-Sep	Black Fast	11	1:42	
28-Sep		12	1:38	
29-Sep		13	1:34	
30-Sep	Full Moon	14	1:30	
1-Oct		15	1:26	
2-Oct		16	1:22	
3-Oct		17	1:18	
4-Oct		18	1:14	
5-Oct		19	1:10	
6-Oct		20	1:06	
7-Oct		21	1:03	
8-Oct		22	12:59	
9-Oct		23	12:55	
10-Oct		24	12:51	
11-Oct		25	12:47	

12-Oct		26	12:43	
13-Oct		27	12:39	
14-Oct		28	12:35	
15-Oct	New Moon	0	12:31	
16-Oct		1	12:27	
17-Oct		2	12:23	
18-Oct		3	12:19	
19-Oct		4	12:15	
20-Oct		5	12:11	
21-Oct		6	12:07	
22-Oct		7	12:03	
23-Oct	Fast Begins	8	12:00 PM	
24-Oct		9	11:56 AM	
25-Oct		10	11:52	
26-Oct	Black Fast	11	11:48	
27-Oct		12	11:44	
28-Oct		13	11:40	
29-Oct	Full Moon	14	11:36	
30-Oct		15	11:32	
31-Oct		16	11:28	Samhain
1-Nov		17	11:24	
2-Nov		18	11:20	
3-Nov		19	11:16	
4-Nov		20	11:12	

Date	Day of Walking	Day of the Moon	Time of Calling	Notes *(Times in Ephemeris Time)*
5-Nov		21	11:08 AM	
6-Nov		22	11:04	
7-Nov		23	11:00	
8-Nov		24	10:57	
9-Nov		25	10:53	
10-Nov		26	10:49	
11-Nov		27	10:45	
12-Nov		28	10:41	
13-Nov	New Moon	0	10:37	Total Solar Eclipse 10:12 PM
14-Nov		1	10:33	
15-Nov		2	10:29	
16-Nov		3	10:25	
17-Nov		4	10:21	
18-Nov		5	10:17	
19-Nov		6	10:13	
20-Nov		7	10:09	
21-Nov		8	10:05	
22-Nov	Fast Begins	9	10:01	
23-Nov		10	9:57	
24-Nov		11	9:54	
25-Nov	Black Fast	12	9:50	

26-Nov		13	9:46	
27-Nov		14	9:42	
28-Nov	Full Moon	15	9:38	Appulse Lunar Eclipse 2:34 PM
29-Nov		16	9:34	
30-Nov		17	9:30	
1-Dec		18	9:26	
2-Dec		19	9:22	
3-Dec		20	9:18	
4-Dec		21	9:14	
5-Dec		22	9:10	
6-Dec		23	9:06	
7-Dec		24	9:02	
8-Dec		25	8:58	
9-Dec		26	8:54	
10-Dec		27	8:50	
11-Dec		28	8:46	
12-Dec		29	8:42	
13-Dec	New Moon	0	8:38	
14-Dec		1	8:34	
15-Dec		2	8:30	
16-Dec		3	8:26	
17-Dec		4	8:22	
18-Dec		5	8:18	
19-Dec		6	8:14	

Date	Day of Walking	Day of the Moon	Time of Calling	Notes *(Times in Ephemeris Time)*
20-Dec		7	8:10 AM	
21-Dec		8	8:06	Winter Solstice 11:13 AM
22-Dec	Fast Begins	9	8:02	
23-Dec		10	7:58	
24-Dec		11	7:54	
25-Dec	Black Fast	12	7:51	
26-Dec		13	7:47	
27-Dec		14	7:43	
28-Dec	Full Moon	15	7:39	
29-Dec		16	7:35	
30-Dec		17	7:31	
31-Dec		18	7:27	
2013 1-Jan		19	7:23	
2-Jan		20	7:19	
3-Jan		21	7:15	
4-Jan		22	7:11	
5-Jan		23	7:07	
6-Jan		24	7:03	
7-Jan		25	6:59	
8-Jan		26	6:55	

9-Jan		27	6:51	
10-Jan		28	6:47	
11-Jan	New Moon	0	6:44	
12-Jan		1	6:40	
13-Jan		2	6:36	
14-Jan		3	6:32	
15-Jan		4	6:28	
16-Jan		5	6:24	
17-Jan		6	6:20	
18-Jan		7	6:16	
19-Jan		8	6:12	
20-Jan		9	6:08	
21-Jan	Fast Begins	10	6:04	
22-Jan		11	6:00	
23-Jan		12	5:56	
24-Jan	Black Fast	13	5:52	
25-Jan		14	5:48	
26-Jan		15	5:45	
27-Jan	Full Moon	16	5:41	
28-Jan		17	5:37	
29-Jan		18	5:33	
30-Jan		19	5:29	
31-Jan		20	5:25	Imbolc
1-Feb		21	5:21	

Date	Day of Walking	Day of the Moon	Time of Calling	Notes (*Times in Ephemeris Time*)
2-Feb		22	5:17 AM	
3-Feb		23	5:13	
4-Feb		24	5:09	
5-Feb		25	5:05	
6-Feb		26	5:01	
7-Feb		27	4:57	
8-Feb		28	4:53	
9-Feb		29	4:49	
10-Feb	New Moon	0	4:45	
11-Feb		1	4:42	
12-Feb		2	4:38	
13-Feb		3	4:34	
14-Feb		4	4:30	
15-Feb		5	4:26	
16-Feb		6	4:22	
17-Feb		7	4:18	
18-Feb		8	4:14	
19-Feb	Fast Begins	9	4:10	
20-Feb		10	4:06	
21-Feb		11	4:02	
22-Feb	Black Fast	12	3:58	

23-Feb		13	3:54
24-Feb		14	3:50
25-Feb	Full Moon	15	3:46
26-Feb		16	3:42
27-Feb		17	3:38
28-Feb		18	3:34
1-Mar		19	3:30
2-Mar		20	3:26
3-Mar		21	3:22
4-Mar		22	3:18
5-Mar		23	3:14
6-Mar		24	3:10
7-Mar		25	3:06
8-Mar		26	3:02
9-Mar		27	2:58
10-Mar		28	2:54
11-Mar	New Moon	0	2:50
12-Mar		1	2:46
13-Mar		2	2:42
14-Mar		3	2:38
15-Mar		4	2:34
16-Mar		5	2:30
17-Mar		6	2:26
18-Mar		7	2:22

Date	Day of Walking	Day of the Moon	Time of Calling	Notes (*Times in Ephemeris Time*)
19-Mar		8	2:18 AM	
20-Mar		9	2:14	Vernal Equinox 11:03 AM
21-Mar	Fast Begins	10	2:10	
22-Mar		11	2:06	
23-Mar		12	2:02	
24-Mar	Black Fast	13	1:58	
25-Mar		14	1:54	
26-Mar		15	1:50	
27-Mar	Full Moon	16	1:46	
28-Mar		17	1:42	
29-Mar		18	1:38	
30-Mar		19	1:34	
31-Mar		20	1:30	
1-Apr		21	1:26	
2-Apr		22	1:22	
3-Apr		23	1:18	
4-Apr		24	1:14	
5-Apr		25	1:10	
6-Apr		26	1:06	
7-Apr		27	1:02	
8-Apr		28	12:58	

9-Apr		29	12:54	
10-Apr	New Moon	0	12:50	
11-Apr		1	12:46	
12-Apr		2	12:42	
13-Apr		3	12:38	
14-Apr		4	12:34	
15-Apr		5	12:30	
16-Apr		6	12:25	
17-Apr		7	12:21	
18-Apr		8	12:17	
19-Apr	Fast Begins	9	12:13	
20-Apr		10	12:09	
21-Apr		11	12:05	
22-Apr	Black Fast	12	12:01 AM	
23-Apr		13	11:57 PM	
24-Apr		14	11:53	
25-Apr	Full Moon	15	11:49	Partial Lunar Eclipse 8:09 PM
26-Apr		16	11:45	
27-Apr		17	11:41	
28-Apr		18	11:37	
29-Apr		19	11:33	
30-Apr		20	11:30	Beltane
1-May		21	11:26	
2-May		22	11:22	

Date	Day of Walking	Day of the Moon	Time of Calling	Notes *(Times in Ephemeris Time)*
3-May		23	11:18 PM	
4-May		24	11:14	
5-May		25	11:10	
6-May		26	11:06	
7-May		27	11:02	
8-May		28	10:58	
9-May		29	10:54	
10-May	New Moon	0	10:50	Annular Solar Eclipse 12:26 AM
11-May		1	10:46	
12-May		2	10:42	
13-May		3	10:38	
14-May		4	10:34	
15-May		5	10:31	
16-May		6	10:27	
17-May		7	10:23	
18-May		8	10:19	
19-May	Fast Begins	9	10:15	
20-May		10	10:11	
21-May		11	10:07	
22-May	Black Fast	12	10:03	
23-May		13	9:59	

24-May		14	9:55	
25-May	Full Moon	15	9:51	Appulse Lunar Eclipse 4:11 AM
26-May		16	9:47	
27-May		17	9:43	
28-May		18	9:39	
29-May		19	9:35	
30-May		20	9:31	
31-May		21	9:28	
1-Jun		22	9:24	
2-Jun		23	9:20	
3-Jun		24	9:16	
4-Jun		25	9:12	
5-Jun		26	9:08	
6-Jun		27	9:04	
7-Jun		28	9:00	
8-Jun	New Moon	0	8:56	
9-Jun		1	8:52	
10-Jun		2	8:48	
11-Jun		3	8:44	
12-Jun		4	8:40	
13-Jun		5	8:36	
14-Jun		6	8:32	
15-Jun		7	8:28	
16-Jun		8	8:25	

Date	Day of Walking	Day of the Moon	Time of Calling	Notes *(Times in Ephemeris Time)*
17-Jun	Fast Begins	9	8:21 PM	
18-Jun		10	8:17	
19-Jun		11	8:13	
20-Jun	Black Fast	12	8:09	
21-Jun		13	8:05	Summer Solstice 5:05 AM
22-Jun		14	8:01	
23-Jun	Full Moon	15	7:57	
24-Jun		16	7:53	
25-Jun		17	7:49	
26-Jun		18	7:45	
27-Jun		19	7:41	
28-Jun		20	7:37	
29-Jun		21	7:33	
30-Jun		22	7:29	
1-Jul		23	7:25	
2-Jul		24	7:22	
3-Jul		25	7:18	
4-Jul		26	7:14	
5-Jul		27	7:10	
6-Jul		28	7:06	
7-Jul		29	7:02	

8-Jul	New Moon	0	6:58
9-Jul		1	6:54
10-Jul		2	6:50
11-Jul		3	6:46
12-Jul		4	6:42
13-Jul		5	6:38
14-Jul		6	6:34
15-Jul		7	6:30
16-Jul	Fast Begins	8	6:26
17-Jul		9	6:22
18-Jul		10	6:19
19-Jul	Black Fast	11	6:15
20-Jul		12	6:11
21-Jul		13	6:07
22-Jul	Full Moon	14	6:03
23-Jul		15	5:59
24-Jul		16	5:55
25-Jul		17	5:51
26-Jul		18	5:47
27-Jul		19	5:43
28-Jul		20	5:39
29-Jul		21	5:35
30-Jul		22	5:31
31-Jul		23	5:27

Date	Day of Walking	Day of the Moon	Time of Calling	Notes (*Times in Ephemeris Time*)
1-Aug		24	5:23 PM	Lammas
2-Aug		25	5:19	
3-Aug		26	5:18	
4-Aug		27	5:14	
5-Aug		28	5:10	
6-Aug	New Moon	0	5:06	
7-Aug		1	5:02	
8-Aug		2	4:58	
9-Aug		3	4:54	
10-Aug		4	4:50	
11-Aug		5	4:46	
12-Aug		6	4:42	
13-Aug		7	4:38	
14-Aug		8	4:34	
15-Aug	Fast Begins	9	4:30	
16-Aug		10	4:26	
17-Aug		11	4:22	
18-Aug	Black Fast	12	4:18	
19-Aug		13	4:16	
20-Aug		14	4:12	
21-Aug	Full Moon	15	4:08	

22-Aug		16	4:04	
23-Aug		17	4:00	
24-Aug		18	3:56	
25-Aug		19	3:52	
26-Aug		20	3:48	
27-Aug		21	3:44	
28-Aug		22	3:40	
29-Aug		23	3:36	
30-Aug		24	3:32	
31-Aug		25	3:28	
1-Sep		26	3:24	
2-Sep		27	3:20	
3-Sep		28	3:16	
4-Sep		29	3:12	
5-Sep	New Moon	0	3:09	
6-Sep		1	3:05	
7-Sep		2	3:01	
8-Sep		3	2:57	
9-Sep		4	2:53	
10-Sep		5	2:49	
11-Sep		6	2:45	
12-Sep		7	2:41	
13-Sep	Fast Begins	8	2:37	
14-Sep		9	2:33	

Date	Day of Walking	Day of the Moon	Time of Calling	Notes *(Times in Ephemeris Time)*
15-Sep		10	2:29 PM	
16-Sep	Black Fast	11	2:25	
17-Sep		12	2:21	
18-Sep		13	2:17	
19-Sep	Full Moon	14	2:13	
20-Sep		15	2:09	
21-Sep		16	2:06	
22-Sep		17	2:02	Autumnal Equinox 8:45 PM
23-Sep		18	1:58	
24-Sep		19	1:54	
25-Sep		20	1:50	
26-Sep		21	1:46	
27-Sep		22	1:42	
28-Sep		23	1:38	
29-Sep		24	1:34	
30-Sep		25	1:30	
1-Oct		26	1:26	
2-Oct		27	1:22	
3-Oct		28	1:18	
4-Oct		29	1:14	
5-Oct	New Moon	0	1:10	

6-Oct		1	1:06	
7-Oct		2	1:03	
8-Oct		3	12:59	
9-Oct		4	12:55	
10-Oct		5	12:51	
11-Oct		6	12:47	
12-Oct	Fast Begins	7	12:43	
13-Oct		8	12:39	
14-Oct		9	12:35	
15-Oct	Black Fast	10	12:31	
16-Oct		11	12:27	
17-Oct		12	12:23	
18-Oct	Full Moon	13	12:19	Appulse Lunar Eclipse 11:51 PM
19-Oct		14	12:15	
20-Oct		15	12:11	
21-Oct		16	12:07	
22-Oct		17	12:03	
23-Oct		18	12:00 PM	
24-Oct		19	11:56 AM	
25-Oct		20	11:52	
26-Oct		21	11:48	
27-Oct		22	11:44	
28-Oct		23	11:40	
29-Oct		24	11:36	

Date	Day of Walking	Day of the Moon	Time of Calling	Notes (*Times in Ephemeris Time*)
30-Oct		25	11:32 AM	
31-Oct		26	11:28	Samhain
1-Nov		27	11:24	
2-Nov		28	11:20	
3-Nov	New Moon	0	11:16	Annular / Total Solar Eclipse 12:47 PM
4-Nov		1	11:12	
5-Nov		2	11:08	
6-Nov		3	11:04	
7-Nov		4	11:00	
8-Nov		5	10:57	
9-Nov		6	10:53	
10-Nov		7	10:49	
11-Nov	Fast Begins	8	10:45	
12-Nov		9	10:41	
13-Nov		10	10:37	
14-Nov	Black Fast	11	10:33	
15-Nov		12	10:29	
16-Nov		13	10:25	
17-Nov	Full Moon	14	10:21	
18-Nov		15	10:17	

19-Nov		16	10:13
20-Nov		17	10:09
21-Nov		18	10:05
22-Nov		19	10:01
23-Nov		20	9:57
24-Nov		21	9:54
25-Nov		22	9:50
26-Nov		23	9:46
27-Nov		24	9:42
28-Nov		25	9:38
29-Nov		26	9:34
30-Nov		27	9:30
1-Dec		28	9:26
2-Dec		29	9:22
3-Dec	New Moon	0	9:18
4-Dec		1	9:14
5-Dec		2	9:10
6-Dec		3	9:06
7-Dec		4	9:02
8-Dec		5	8:58
9-Dec		6	8:54
10-Dec		7	8:50
11-Dec	Fast Begins	8	8:46
12-Dec		9	8:42

Date	Day of Walking	Day of the Moon	Time of Calling	Notes *(Times in Ephemeris Time)*
13-Dec		10	8:38 AM	
14-Dec	Black Fast	11	8:34	
15-Dec		12	8:30	
16-Dec		13	8:26	
17-Dec	Full Moon	14	8:22	
18-Dec		15	8:18	
19-Dec		16	8:14	
20-Dec		17	8:10	
21-Dec		18	8:06	Winter Solstice 5:12 PM
22-Dec		19	8:02	
23-Dec		20	7:58	
24-Dec		21	7:54	
25-Dec		22	7:51	
26-Dec		23	7:47	
27-Dec		24	7:43	
28-Dec		25	7:39	
29-Dec		26	7:35	
30-Dec		27	7:31	
31-Dec		28	7:27	
2014 1-Jan	New Moon	0	7:23	

2-Jan		1	7:19
3-Jan		2	7:15
4-Jan		3	7:11
5-Jan		4	7:07
6-Jan		5	7:03
7-Jan		6	6:59
8-Jan		7	6:55
9-Jan		8	6:51
10-Jan	Fast Begins	9	6:47
11-Jan		10	6:44
12-Jan		11	6:40
13-Jan	Black Fast	12	6:36
14-Jan		13	6:32
15-Jan		14	6:28
16-Jan	Full Moon	15	6:24
17-Jan		16	6:20
18-Jan		17	6:16
19-Jan		18	6:12
20-Jan		19	6:08
21-Jan		20	6:04
22-Jan		21	6:00
23-Jan		22	5:56
24-Jan		23	5:52
25-Jan		24	5:48

Date	Day of Walking	Day of the Moon	Time of Calling	Notes *(Times in Ephemeris Time)*
26-Jan		25	5:45 AM	
27-Jan		26	5:41	
28-Jan		27	5:37	
29-Jan		28	5:33	
30-Jan	New Moon	0	5:29	
31-Jan		1	5:25	Imbolc
1-Feb		2	5:21	
2-Feb		3	5:17	
3-Feb		4	5:13	
4-Feb		5	5:09	
5-Feb		6	5:05	
6-Feb		7	5:01	
7-Feb		8	4:57	
8-Feb	Fast Begins	9	4:53	
9-Feb		10	4:49	
10-Feb		11	4:45	
11-Feb	Black Fast	12	4:42	
12-Feb		13	4:38	
13-Feb		14	4:34	
14-Feb	Full Moon	15	4:30	
15-Feb		16	4:26	

16-Feb		17	4:22	
17-Feb		18	4:18	
18-Feb		19	4:14	
19-Feb		20	4:10	
20-Feb		21	4:06	
21-Feb		22	4:02	
22-Feb		23	3:58	
23-Feb		24	3:54	
24-Feb		25	3:50	
25-Feb		26	3:46	
26-Feb		27	3:42	
27-Feb		28	3:38	
28-Feb		29	3:34	
1-Mar	New Moon	0	3:30	
2-Mar		1	3:26	
3-Mar		2	3:22	
4-Mar		3	3:18	
5-Mar		4	3:14	
6-Mar		5	3:10	
7-Mar		6	3:06	
8-Mar		7	3:02	
9-Mar		8	2:58	
10-Mar	Fast Begins	9	2:54	
11-Mar		10	2:50	

Date	Day of Walking	Day of the Moon	Time of Calling	Notes (*Times in Ephemeris Time*)
12-Mar		11	2:46 AM	
13-Mar	Black Fast	12	2:42	
14-Mar		13	2:38	
15-Mar		14	2:34	
16-Mar	Full Moon	15	2:30	
17-Mar		16	2:26	
18-Mar		17	2:22	
19-Mar		18	2:18	
20-Mar		19	2:14	Vernal Equinox 4:58 PM
21-Mar		20	2:10	
22-Mar		21	2:06	
23-Mar		22	2:02	
24-Mar		23	1:58	
25-Mar		24	1:54	
26-Mar		25	1:50	
27-Mar		26	1:46	
28-Mar		27	1:42	
29-Mar		28	1:38	
30-Mar	New Moon	0	1:34	
31-Mar		1	1:30	
1-Apr		2	1:26	

2-Apr		3	1:22	
3-Apr		4	1:18	
4-Apr		5	1:14	
5-Apr		6	1:10	
6-Apr		7	1:06	
7-Apr		8	1:02	
8-Apr		9	12:58	
9-Apr	Fast Begins	10	12:54	
10-Apr		11	12:50	
11-Apr		12	12:46	
12-Apr	Black Fast	13	12:42	
13-Apr		14	12:38	
14-Apr		15	12:34	
15-Apr	Full Moon	16	12:30	Total Lunar Eclipse 7:47 AM
16-Apr		17	12:25	
17-Apr		18	12:21	
18-Apr		19	12:17	
19-Apr		20	12:13	
20-Apr		21	12:09	
21-Apr		22	12:05	
22-Apr		23	12:01 AM	
23-Apr		24	11:57 PM	
24-Apr		25	11:53	
25-Apr		26	11:49	

Date	Day of Walking	Day of the Moon	Time of Calling	Notes (*Times in Ephemeris Time*)
26-Apr		27	11:45 PM	
27-Apr		28	11:41	
28-Apr		29	11:37	
29-Apr	New Moon	0	11:33	Annular non-C Solar Eclipse 6:04 AM
30-Apr		1	11:30	Beltane
1-May		2	11:26	
2-May		3	11:22	
3-May		4	11:18	
4-May		5	11:14	
5-May		6	11:10	
6-May		7	11:06	
7-May		8	11:02	
8-May	Fast Begins	9	10:58	
9-May		10	10:54	
10-May		11	10:50	
11-May	Black Fast	12	10:46	
12-May		13	10:42	
13-May		14	10:38	
14-May	Full Moon	15	10:34	
15-May		16	10:31	

16-May		17	10:27
17-May		18	10:23
18-May		19	10:19
19-May		20	10:15
20-May		21	10:11
21-May		22	10:07
22-May		23	10:03
23-May		24	9:59
24-May		25	9:55
25-May		26	9:51
26-May		27	9:47
27-May		28	9:43
28-May	New Moon	0	9:39
29-May		1	9:35
30-May		2	9:31
31-May		3	9:28
1-Jun		4	9:24
2-Jun		5	9:20
3-Jun		6	9:16
4-Jun		7	9:12
5-Jun		8	9:08
6-Jun		9	9:04
7-Jun	Fast Begins	10	9:00
8-Jun		11	8:56

Date	Day of Walking	Day of the Moon	Time of Calling	Notes *(Times in Ephemeris Time)*
9-Jun		12	8:52 PM	
10-Jun	Black Fast	13	8:48	
11-Jun		14	8:44	
12-Jun		15	8:40	
13-Jun	Full Moon	16	8:36	
14-Jun		17	8:32	
15-Jun		18	8:28	
16-Jun		19	8:25	
17-Jun		20	8:21	
18-Jun		21	8:17	
19-Jun		22	8:13	
20-Jun		23	8:09	
21-Jun		24	8:05	Summer Solstice 10:52 AM
22-Jun		25	8:01	
23-Jun		26	7:57	
24-Jun		27	7:53	
25-Jun		28	7:49	
26-Jun		29	7:45	
27-Jun	New Moon	0	7:41	
28-Jun		1	7:37	
29-Jun		2	7:33	

30-Jun		3	7:29
1-Jul		4	7:25
2-Jul		5	7:22
3-Jul		6	7:18
4-Jul		7	7:14
5-Jul		8	7:10
6-Jul	Fast Begins	9	7:06
7-Jul		10	7:02
8-Jul		11	6:58
9-Jul	Black Fast	12	6:54
10-Jul		13	6:50
11-Jul		14	6:46
12-Jul	Full Moon	15	6:42
13-Jul		16	6:38
14-Jul		17	6:34
15-Jul		18	6:30
16-Jul		19	6:26
17-Jul		20	6:22
18-Jul		21	6:19
19-Jul		22	6:15
20-Jul		23	6:11
21-Jul		24	6:07
22-Jul		25	6:03
23-Jul		26	5:59

Date	Day of Walking	Day of the Moon	Time of Calling	Notes *(Times in Ephemeris Time)*
24-Jul		27	5:55 PM	
25-Jul		28	5:51	
26-Jul	New Moon	0	5:47	
27-Jul		1	5:43	
28-Jul		2	5:39	
29-Jul		3	5:35	
30-Jul		4	5:31	
31-Jul		5	5:27	
1-Aug		6	5:23	Lammas
2-Aug		7	5:19	
3-Aug		8	5:18	
4-Aug	Fast Begins	9	5:14	
5-Aug		10	5:10	
6-Aug		11	5:06	
7-Aug	Black Fast	12	5:02	
8-Aug		13	4:58	
9-Aug		14	4:54	
10-Aug	Full Moon	15	4:50	
11-Aug		16	4:46	
12-Aug		17	4:42	
13-Aug		18	4:38	

14-Aug		19	4:34
15-Aug		20	4:30
16-Aug		21	4:26
17-Aug		22	4:22
18-Aug		23	4:18
19-Aug		24	4:16
20-Aug		25	4:12
21-Aug		26	4:08
22-Aug		27	4:04
23-Aug		28	4:00
24-Aug		29	3:56
25-Aug	New Moon	0	3:52
26-Aug		1	3:48
27-Aug		2	3:44
28-Aug		3	3:40
29-Aug		4	3:36
30-Aug		5	3:32
31-Aug		6	3:28
1-Sep		7	3:24
2-Sep		8	3:20
3-Sep	Fast Begins	9	3:16
4-Sep		10	3:12
5-Sep		11	3:09
6-Sep	Black Fast	12	3:05

Date	Day of Walking	Day of the Moon	Time of Calling	Notes *(Times in Ephemeris Time)*
7-Sep		13	3:01 PM	
8-Sep		14	2:57	
9-Sep	Full Moon	15	2:53	
10-Sep		16	2:49	
11-Sep		17	2:45	
12-Sep		18	2:41	
13-Sep		19	2:37	
14-Sep		20	2:33	
15-Sep		21	2:29	
16-Sep		22	2:25	
17-Sep		23	2:21	
18-Sep		24	2:17	
19-Sep		25	2:13	
20-Sep		26	2:09	
21-Sep		27	2:06	
22-Sep		28	2:02	
23-Sep		29	1:58	Autumnal Equinox 2:30 AM
24-Sep	New Moon	0	1:54	
25-Sep		1	1:50	
26-Sep		2	1:46	
27-Sep		3	1:42	

28-Sep		4	1:38	
29-Sep		5	1:34	
30-Sep		6	1:30	
1-Oct		7	1:26	
2-Oct	Fast Begins	8	1:22	
3-Oct		9	1:18	
4-Oct		10	1:14	
5-Oct	Black Fast	11	1:10	
6-Oct		12	1:06	
7-Oct		13	1:03	
8-Oct	Full Moon	14	12:59	Total Lunar Eclipse 10:56 AM
9-Oct		15	12:55	
10-Oct		16	12:51	
11-Oct		17	12:47	
12-Oct		18	12:43	
13-Oct		19	12:39	
14-Oct		20	12:35	
15-Oct		21	12:31	
16-Oct		22	12:27	
17-Oct		23	12:23	
18-Oct		24	12:19	
19-Oct		25	12:15	
20-Oct		26	12:11	
21-Oct		27	12:07	

Date	Day of Walking	Day of the Moon	Time of Calling	Notes *(Times in Ephemeris Time)*
22-Oct		28	12:03 PM	
23-Oct	New Moon	0	12:00 PM	Partial Solar Eclipse 9:45 PM
24-Oct		1	11:56 AM	
25-Oct		2	11:52	
26-Oct		3	11:48	
27-Oct		4	11:44	
28-Oct		5	11:40	
29-Oct		6	11:36	
30-Oct		7	11:32	
31-Oct	Fast Begins	8	11:28	Samhain
1-Nov		9	11:24	
2-Nov		10	11:20	
3-Nov	Black Fast	11	11:16	
4-Nov		12	11:12	
5-Nov		13	11:08	
6-Nov	Full Moon	14	11:04	
7-Nov		15	11:00	
8-Nov		16	10:57	
9-Nov		17	10:53	
10-Nov		18	10:49	
11-Nov		19	10:45	

12-Nov		20	10:41
13-Nov		21	10:37
14-Nov		22	10:33
15-Nov		23	10:29
16-Nov		24	10:25
17-Nov		25	10:21
18-Nov		26	10:17
19-Nov		27	10:13
20-Nov		28	10:09
21-Nov		29	10:05
22-Nov	New Moon	0	10:01
23-Nov		1	9:57
24-Nov		2	9:54
25-Nov		3	9:50
26-Nov		4	9:46
27-Nov		5	9:42
28-Nov		6	9:38
29-Nov		7	9:34
30-Nov	Fast Begins	8	9:30
1-Dec		9	9:26
2-Dec		10	9:22
3-Dec	Black Fast	11	9:18
4-Dec		12	9:14
5-Dec		13	9:10

Date	Day of Walking	Day of the Moon	Time of Calling	Notes *(Times in Ephemeris Time)*
6-Dec	Full Moon	14	9:06 AM	
7-Dec		15	9:02	
8-Dec		16	8:58	
9-Dec		17	8:54	
10-Dec		18	8:50	
11-Dec		19	8:46	
12-Dec		20	8:42	
13-Dec		21	8:38	
14-Dec		22	8:34	
15-Dec		23	8:30	
16-Dec		24	8:26	
17-Dec		25	8:22	
18-Dec		26	8:18	
19-Dec		27	8:14	
20-Dec		28	8:10	
21-Dec		29	8:06	Winter Solstice 11:04 PM
22-Dec	New Moon	0	8:02	
23-Dec		1	7:58	
24-Dec		2	7:54	
25-Dec		3	7:51	
26-Dec		4	7:47	

	27-Dec		5	7:43
	28-Dec		6	7:39
	29-Dec		7	7:35
	30-Dec	Fast Begins	8	7:31
	31-Dec		9	7:27
2015	1-Jan		10	7:23
	2-Jan	Black Fast	11	7:19
	3-Jan		12	7:15
	4-Jan		13	7:11
	5-Jan	Full Moon	14	7:07
	6-Jan		15	7:03
	7-Jan		16	6:59
	8-Jan		17	6:55
	9-Jan		18	6:51
	10-Jan		19	6:47
	11-Jan		20	6:44
	12-Jan		21	6:40
	13-Jan		22	6:36
	14-Jan		23	6:32
	15-Jan		24	6:28
	16-Jan		25	6:24
	17-Jan		26	6:20
	18-Jan		27	6:16

Date	Day of Walking	Day of the Moon	Time of Calling	Notes (*Times in Ephemeris Time*)
19-Jan		28	6:12 AM	
20-Jan	New Moon	0	6:08	
21-Jan		1	6:04	
22-Jan		2	6:00	
23-Jan		3	5:56	
24-Jan		4	5:52	
25-Jan		5	5:48	
26-Jan		6	5:45	
27-Jan		7	5:41	
28-Jan	Fast Begins	8	5:37	
29-Jan		9	5:33	
30-Jan		10	5:29	
31-Jan	Black Fast	11	5:25	Imbolc
1-Feb		12	5:21	
2-Feb		13	5:17	
3-Feb	Full Moon	14	5:13	
4-Feb		15	5:09	
5-Feb		16	5:05	
6-Feb		17	5:01	
7-Feb		18	4:57	
8-Feb		19	4:53	

9-Feb		20	4:49
10-Feb		21	4:45
11-Feb		22	4:42
12-Feb		23	4:38
13-Feb		24	4:34
14-Feb		25	4:30
15-Feb		26	4:26
16-Feb		27	4:22
17-Feb		28	4:18
18-Feb	New Moon	0	4:14
19-Feb		1	4:10
20-Feb		2	4:06
21-Feb		3	4:02
22-Feb		4	3:58
23-Feb		5	3:54
24-Feb		6	3:50
25-Feb		7	3:46
26-Feb		8	3:42
27-Feb	Fast Begins	9	3:38
28-Feb		10	3:34
1-Mar		11	3:30
2-Mar	Black Fast	12	3:26
3-Mar		13	3:22
4-Mar		14	3:18

Date	Day of Walking	Day of the Moon	Time of Calling	Notes *(Times in Ephemeris Time)*
5-Mar	Full Moon	15	3:14 AM	
6-Mar		16	3:10	
7-Mar		17	3:06	
8-Mar		18	3:02	
9-Mar		19	2:58	
10-Mar		20	2:54	
11-Mar		21	2:50	
12-Mar		22	2:46	
13-Mar		23	2:42	
14-Mar		24	2:38	
15-Mar		25	2:34	
16-Mar		26	2:30	
17-Mar		27	2:26	
18-Mar		28	2:22	
19-Mar		29	2:18	
20-Mar	New Moon	0	2:14	Total Solar Eclipse 9:46 AM / Vernal Equinox 10:46 PM
21-Mar		1	2:10	
22-Mar		2	2:06	
23-Mar		3	2:02	
24-Mar		4	1:58	

25-Mar		5	1:54	
26-Mar		6	1:50	
27-Mar		7	1:46	
28-Mar		8	1:42	
29-Mar	Fast Begins	9	1:38	
30-Mar		10	1:34	
31-Mar		11	1:30	
1-Apr	Black Fast	12	1:26	
2-Apr		13	1:22	
3-Apr		14	1:18	
4-Apr	Full Moon	15	1:14	Total Lunar Eclipse 12:01 PM
5-Apr		16	1:10	
6-Apr		17	1:06	
7-Apr		18	1:02	
8-Apr		19	12:58	
9-Apr		20	12:54	
10-Apr		21	12:50	
11-Apr		22	12:46	
12-Apr		23	12:42	
13-Apr		24	12:38	
14-Apr		25	12:34	
15-Apr		26	12:30	
16-Apr		27	12:25	
17-Apr		28	12:21	

Date	Day of Walking	Day of the Moon	Time of Calling	Notes *(Times in Ephemeris Time)*
18-Apr	New Moon	0	12:17 AM	
19-Apr		1	12:13	
20-Apr		2	12:09	
21-Apr		3	12:05	
22-Apr		4	12:01 AM	
23-Apr		5	11:57 PM	
24-Apr		6	11:53	
25-Apr		7	11:49	
26-Apr		8	11:45	
27-Apr		9	11:41	
28-Apr	Fast Begins	10	11:37	
29-Apr		11	11:33	
30-Apr		12	11:30	Beltane
1-May	Black Fast	13	11:26	
2-May		14	11:22	
3-May		15	11:18	
4-May	Full Moon	16	11:14	
5-May		17	11:10	
6-May		18	11:06	
7-May		19	11:02	
8-May		20	10:58	

9-May		21	10:54	
10-May		22	10:50	
11-May		23	10:46	
12-May		24	10:42	
13-May		25	10:38	
14-May		26	10:34	
15-May		27	10:31	
16-May		28	10:27	
17-May		29	10:23	
18-May	New Moon	0	10:19	
19-May		1	10:15	
20-May		2	10:11	
21-May		3	10:07	
22-May		4	10:03	
23-May		5	9:59	
24-May		6	9:55	
25-May		7	9:51	
26-May		8	9:47	
27-May	Fast Begins	9	9:43	
28-May		10	9:39	
29-May		11	9:35	
30-May	Black Fast	12	9:31	
31-May		13	9:28	
1-Jun		14	9:24	

Date	Day of Walking	Day of the Moon	Time of Calling	Notes *(Times in Ephemeris Time)*
2-Jun	Full Moon	15	9:20 PM	
3-Jun		16	9:16	
4-Jun		17	9:12	
5-Jun		18	9:08	
6-Jun		19	9:04	
7-Jun		20	9:00	
8-Jun		21	8:56	
9-Jun		22	8:52	
10-Jun		23	8:48	
11-Jun		24	8:44	
12-Jun		25	8:40	
13-Jun		26	8:36	
14-Jun		27	8:32	
15-Jun		28	8:28	
16-Jun	New Moon	0	8:25	
17-Jun		1	8:21	
18-Jun		2	8:17	
19-Jun		3	8:13	
20-Jun		4	8:09	
21-Jun		5	8:05	Summer Solstice 4:39 PM
22-Jun		6	8:01	

23-Jun		7	7:57
24-Jun		8	7:53
25-Jun		9	7:49
26-Jun	Fast Begins	10	7:45
27-Jun		11	7:41
28-Jun		12	7:37
29-Jun	Black Fast	13	7:33
30-Jun		14	7:29
1-Jul		15	7:25
2-Jul	Full Moon	16	7:22
3-Jul		17	7:18
4-Jul		18	7:14
5-Jul		19	7:10
6-Jul		20	7:06
7-Jul		21	7:02
8-Jul		22	6:58
9-Jul		23	6:54
10-Jul		24	6:50
11-Jul		25	6:46
12-Jul		26	6:42
13-Jul		27	6:38
14-Jul		28	6:34
15-Jul		29	6:30
16-Jul	New Moon	0	6:26

Date	Day of Walking	Day of the Moon	Time of Calling	Notes *(Times in Ephemeris Time)*
17-Jul		1	6:22 PM	
18-Jul		2	6:19	
19-Jul		3	6:15	
20-Jul		4	6:11	
21-Jul		5	6:07	
22-Jul		6	6:03	
23-Jul		7	5:59	
24-Jul		8	5:55	
25-Jul	Fast Begins	9	5:51	
26-Jul		10	5:47	
27-Jul		11	5:43	
28-Jul	Black Fast	12	5:39	
29-Jul		13	5:35	
30-Jul		14	5:31	
31-Jul	Full Moon	15	5:27	
1-Aug		16	5:23	Lammas
2-Aug		17	5:19	
3-Aug		18	5:18	
4-Aug		19	5:14	
5-Aug		20	5:10	
6-Aug		21	5:06	

7-Aug		22	5:02
8-Aug		23	4:58
9-Aug		24	4:54
10-Aug		25	4:50
11-Aug		26	4:46
12-Aug		27	4:42
13-Aug		28	4:38
14-Aug	New Moon	0	4:34
15-Aug		1	4:30
16-Aug		2	4:26
17-Aug		3	4:22
18-Aug		4	4:18
19-Aug		5	4:16
20-Aug		6	4:12
21-Aug		7	4:08
22-Aug		8	4:04
23-Aug	Fast Begins	9	4:00
24-Aug		10	3:56
25-Aug		11	3:52
26-Aug	Black Fast	12	3:48
27-Aug		13	3:44
28-Aug		14	3:40
29-Aug	Full Moon	15	3:36
30-Aug		16	3:32

Date	Day of Walking	Day of the Moon	Time of Calling	Notes (*Times in Ephemeris Time*)
31-Aug		17	3:28 PM	
1-Sep		18	3:24	
2-Sep		19	3:20	
3-Sep		20	3:16	
4-Sep		21	3:12	
5-Sep		22	3:09	
6-Sep		23	3:05	
7-Sep		24	3:01	
8-Sep		25	2:57	
9-Sep		26	2:53	
10-Sep		27	2:49	
11-Sep		28	2:45	
12-Sep		29	2:41	
13-Sep	New Moon	0	2:37	Partial Solar Eclipse 6:55 AM
14-Sep		1	2:33	
15-Sep		2	2:29	
16-Sep		3	2:25	
17-Sep		4	2:21	
18-Sep		5	2:17	
19-Sep		6	2:13	
20-Sep		7	2:09	

21-Sep		8	2:06	
22-Sep	Fast Begins	9	2:02	
23-Sep		10	1:58	Autumnal Equinox 8:22 AM
24-Sep		11	1:54	
25-Sep	Black Fast	12	1:50	
26-Sep		13	1:46	
27-Sep		14	1:42	
28-Sep	Full Moon	15	1:38	Total Lunar Eclipse 2:48 AM
29-Sep		16	1:34	
30-Sep		17	1:30	
1-Oct		18	1:26	
2-Oct		19	1:22	
3-Oct		20	1:18	
4-Oct		21	1:14	
5-Oct		22	1:10	
6-Oct		23	1:06	
7-Oct		24	1:03	
8-Oct		25	12:59	
9-Oct		26	12:55	
10-Oct		27	12:51	
11-Oct		28	12:47	
12-Oct		29	12:43	
13-Oct	New Moon	0	12:39	
14-Oct		1	12:35	

Date	Day of Walking	Day of the Moon	Time of Calling	Notes *(Times in Ephemeris Time)*
15-Oct		2	12:31 PM	
16-Oct		3	12:27	
17-Oct		4	12:23	
18-Oct		5	12:19	
19-Oct		6	12:15	
20-Oct		7	12:11	
21-Oct	Fast Begins	8	12:07	
22-Oct		9	12:03	
23-Oct		10	12:00 PM	
24-Oct	Black Fast	11	11:56 AM	
25-Oct		12	11:52	
26-Oct		13	11:48	
27-Oct	Full Moon	14	11:44	
28-Oct		15	11:40	
29-Oct		16	11:36	
30-Oct		17	11:32	
31-Oct		18	11:28	Samhain
1-Nov		19	11:24	
2-Nov		20	11:20	
3-Nov		21	11:16	
4-Nov		22	11:12	

5-Nov		23	11:08
6-Nov		24	11:04
7-Nov		25	11:00
8-Nov		26	10:57
9-Nov		27	10:53
10-Nov		28	10:49
11-Nov	New Moon	0	10:45
12-Nov		1	10:41
13-Nov		2	10:37
14-Nov		3	10:33
15-Nov		4	10:29
16-Nov		5	10:25
17-Nov		6	10:21
18-Nov		7	10:17
19-Nov	Fast Begins	8	10:13
20-Nov		9	10:09
21-Nov		10	10:05
22-Nov	Black Fast	11	10:01
23-Nov		12	9:57
24-Nov		13	9:54
25-Nov	Full Moon	14	9:50
26-Nov		15	9:46
27-Nov		16	9:42
28-Nov		17	9:38

Date	Day of Walking	Day of the Moon	Time of Calling	Notes (*Times in Ephemeris Time*)
29-Nov		18	9:34 AM	
30-Nov		19	9:30	
1-Dec		20	9:26	
2-Dec		21	9:22	
3-Dec		22	9:18	
4-Dec		23	9:14	
5-Dec		24	9:10	
6-Dec		25	9:06	
7-Dec		26	9:02	
8-Dec		27	8:58	
9-Dec		28	8:54	
10-Dec		29	8:50	
11-Dec	New Moon	0	8:46	
12-Dec		1	8:42	
13-Dec		2	8:38	
14-Dec		3	8:34	
15-Dec		4	8:30	
16-Dec		5	8:26	
17-Dec		6	8:22	
18-Dec	Fast Begins	7	8:18	
19-Dec		8	8:14	

	20-Dec		9	8:10	
	21-Dec	Black Fast	10	8:06	
	22-Dec		11	8:02	Winter Solstice 4:49 AM
	23-Dec		12	7:58	
	24-Dec	Full Moon	13	7:54	
	25-Dec		14	7:51	
	26-Dec		15	7:47	
	27-Dec		16	7:43	
	28-Dec		17	7:39	
	29-Dec		18	7:35	
	30-Dec		19	7:31	
	31-Dec		20	7:27	
2016	1-Jan		21	7:23	
	2-Jan		22	7:19	
	3-Jan		23	7:15	
	4-Jan		24	7:11	
	5-Jan		25	7:07	
	6-Jan		26	7:03	
	7-Jan		27	6:59	
	8-Jan		28	6:55	
	9-Jan		29	6:51	
	10-Jan	New Moon	0	6:47	
	11-Jan		1	6:44	

Date	Day of Walking	Day of the Moon	Time of Calling	Notes (*Times in Ephemeris Time*)
12-Jan		2	6:40 AM	
13-Jan		3	6:36	
14-Jan		4	6:32	
15-Jan		5	6:28	
16-Jan		6	6:24	
17-Jan		7	6:20	
18-Jan	Fast Begins	8	6:16	
19-Jan		9	6:12	
20-Jan		10	6:08	
21-Jan	Black Fast	11	6:04	
22-Jan		12	6:00	
23-Jan		13	5:56	
24-Jan	Full Moon	14	5:52	
25-Jan		15	5:48	
26-Jan		16	5:45	
27-Jan		17	5:41	
28-Jan		18	5:37	
29-Jan		19	5:33	
30-Jan		20	5:29	
31-Jan		21	5:25	Imbolc
1-Feb		22	5:21	

2-Feb		23	5:17
3-Feb		24	5:13
4-Feb		25	5:09
5-Feb		26	5:05
6-Feb		27	5:01
7-Feb		28	4:57
8-Feb	New Moon	0	4:53
9-Feb		1	4:49
10-Feb		2	4:45
11-Feb		3	4:42
12-Feb		4	4:38
13-Feb		5	4:34
14-Feb		6	4:30
15-Feb		7	4:26
16-Feb	Fast Begins	8	4:22
17-Feb		9	4:18
18-Feb		10	4:14
19-Feb	Black Fast	11	4:10
20-Feb		12	4:06
21-Feb		13	4:02
22-Feb	Full Moon	14	3:58
23-Feb		15	3:54
24-Feb		16	3:50
25-Feb		17	3:46

Date	Day of Walking	Day of the Moon	Time of Calling	Notes *(Times in Ephemeris Time)*
26-Feb		18	3:42 AM	
27-Feb		19	3:38	
28-Feb		20	3:34	
29-Feb		21	3:30	
1-Mar		22	3:26	
2-Mar		23	3:22	
3-Mar		24	3:18	
4-Mar		25	3:14	
5-Mar		26	3:10	
6-Mar		27	3:06	
7-Mar		28	3:02	
8-Mar		29	2:58	
9-Mar	New Moon	0	2:54	Total Solar Eclipse 1:58 AM
10-Mar		1	2:50	
11-Mar		2	2:46	
12-Mar		3	2:42	
13-Mar		4	2:38	
14-Mar		5	2:35	
15-Mar		6	2:32	
16-Mar		7	2:30	
17-Mar	Fast Begins	8	2:26	

18-Mar		9	2:22	
19-Mar		10	2:18	
20-Mar	Black Fast	11	2:14	Vernal Equinox 4:31 AM
21-Mar		12	2:10	
22-Mar		13	2:06	
23-Mar	Full Moon	14	2:02	Appulse Lunar Eclipse 11:48 AM
24-Mar		15	1:58	
25-Mar		16	1:54	
26-Mar		17	1:50	
27-Mar		18	1:46	
28-Mar		19	1:42	
29-Mar		20	1:38	
30-Mar		21	1:34	
31-Mar		22	1:30	
1-Apr		23	1:26	
2-Apr		24	1:22	
3-Apr		25	1:18	
4-Apr		26	1:14	
5-Apr		27	1:10	
6-Apr		28	1:06	
7-Apr	New Moon	0	1:02	
8-Apr		1	12:58	
9-Apr		2	12:54	
10-Apr		3	12:50	

Date	Day of Walking	Day of the Moon	Time of Calling	Notes (*Times in Ephemeris Time*)
11-Apr		4	12:46 AM	
12-Apr		5	12:42	
13-Apr		6	12:38	
14-Apr		7	12:34	
15-Apr		8	12:30	
16-Apr	Fast Begins	9	12:25	
17-Apr		10	12:21	
18-Apr		11	12:17	
19-Apr	Black Fast	12	12:13	
20-Apr		13	12:09	
21-Apr		14	12:05	
22-Apr	Full Moon	15	12:01 AM	
23-Apr		16	11:57 PM	
24-Apr		17	11:53	
25-Apr		18	11:49	
26-Apr		19	11:45	
27-Apr		20	11:41	
28-Apr		21	11:37	
29-Apr		22	11:33	
30-Apr		23	11:30	Beltane
1-May		24	11:26	

2-May		25	11:22
3-May		26	11:18
4-May		27	11:14
5-May		28	11:10
6-May	New Moon	0	11:06
7-May		1	11:02
8-May		2	10:58
9-May		3	10:54
10-May		4	10:50
11-May		5	10:46
12-May		6	10:42
13-May		7	10:38
14-May		8	10:34
15-May	Fast Begins	9	10:31
16-May		10	10:27
17-May		11	10:23
18-May	Black Fast	12	10:19
19-May		13	10:15
20-May		14	10:11
21-May	Full Moon	15	10:07
22-May		16	10:03
23-May		17	9:59
24-May		18	9:55
25-May		19	9:51

Date	Day of Walking	Day of the Moon	Time of Calling	Notes (*Times in Ephemeris Time*)
26-May		20	9:47 PM	
27-May		21	9:43	
28-May		22	9:39	
29-May		23	9:35	
30-May		24	9:31	
31-May		25	9:28	
1-Jun		26	9:24	
2-Jun		27	9:20	
3-Jun		28	9:16	
4-Jun		29	9:12	
5-Jun	New Moon	0	9:08	
6-Jun		1	9:04	
7-Jun		2	9:00	
8-Jun		3	8:56	
9-Jun		4	8:52	
10-Jun		5	8:48	
11-Jun		6	8:44	
12-Jun		7	8:40	
13-Jun		8	8:36	
14-Jun	Fast Begins	9	8:32	
15-Jun		10	8:28	

16-Jun		11	8:25	
17-Jun	Black Fast	12	8:21	
18-Jun		13	8:17	
19-Jun		14	8:13	
20-Jun	Full Moon	15	8:09	Summer Solstice 10:35 PM
21-Jun		16	8:05	
22-Jun		17	8:01	
23-Jun		18	7:57	
24-Jun		19	7:53	
25-Jun		20	7:49	
26-Jun		21	7:45	
27-Jun		22	7:41	
28-Jun		23	7:37	
29-Jun		24	7:33	
30-Jun		25	7:29	
1-Jul		26	7:25	
2-Jul		27	7:22	
3-Jul		28	7:18	
4-Jul	New Moon	0	7:14	
5-Jul		1	7:10	
6-Jul		2	7:06	
7-Jul		3	7:02	
8-Jul		4	6:58	
9-Jul		5	6:54	

Date	Day of Walking	Day of the Moon	Time of Calling	Notes (*Times in Ephemeris Time*)
10-Jul		6	6:50 PM	
11-Jul		7	6:46	
12-Jul		8	6:42	
13-Jul	Fast Begins	9	6:38	
14-Jul		10	6:34	
15-Jul		11	6:30	
16-Jul	Black Fast	12	6:26	
17-Jul		13	6:22	
18-Jul		14	6:19	
19-Jul	Full Moon	15	6:15	
20-Jul		16	6:11	
21-Jul		17	6:07	
22-Jul		18	6:03	
23-Jul		19	5:59	
24-Jul		20	5:55	
25-Jul		21	5:51	
26-Jul		22	5:47	
27-Jul		23	5:43	
28-Jul		24	5:39	
29-Jul		25	5:35	
30-Jul		26	5:31	

31-Jul		27	5:27	
1-Aug		28	5:23	Lammas
2-Aug	New Moon	0	5:19	
3-Aug		1	5:18	
4-Aug		2	5:14	
5-Aug		3	5:10	
6-Aug		4	5:06	
7-Aug		5	5:02	
8-Aug		6	4:58	
9-Aug		7	4:54	
10-Aug		8	4:50	
11-Aug		9	4:46	
12-Aug	Fast Begins	10	4:42	
13-Aug		11	4:38	
14-Aug		12	4:34	
15-Aug	Black Fast	13	4:30	
16-Aug		14	4:26	
17-Aug		15	4:22	
18-Aug	Full Moon	16	4:18	
19-Aug		17	4:16	
20-Aug		18	4:12	
21-Aug		19	4:08	
22-Aug		20	4:04	
23-Aug		21	4:00	

Date	Day of Walking	Day of the Moon	Time of Calling	Notes (*Times in Ephemeris Time*)
24-Aug		22	3:56 PM	
25-Aug		23	3:52	
26-Aug		24	3:48	
27-Aug		25	3:44	
28-Aug		26	3:40	
29-Aug		27	3:36	
30-Aug		28	3:32	
31-Aug		29	3:28	
1-Sep	New Moon	0	3:24	Annular Solar Eclipse 9:08 AM
2-Sep		1	3:20	
3-Sep		2	3:16	
4-Sep		3	3:12	
5-Sep		4	3:09	
6-Sep		5	3:05	
7-Sep		6	3:01	
8-Sep		7	2:57	
9-Sep		8	2:53	
10-Sep	Fast Begins	9	2:49	
11-Sep		10	2:45	
12-Sep		11	2:41	
13-Sep	Black Fast	12	2:37	

14-Sep		13	2:33	
15-Sep		14	2:29	
16-Sep	Full Moon	15	2:25	Appulse Lunar Eclipse 6:55 PM
17-Sep		16	2:21	
18-Sep		17	2:17	
19-Sep		18	2:13	
20-Sep		19	2:09	
21-Sep		20	2:06	
22-Sep		21	2:02	Autumnal Equinox 2:52 PM
23-Sep		22	1:58	
24-Sep		23	1:54	
25-Sep		24	1:50	
26-Sep		25	1:46	
27-Sep		26	1:42	
28-Sep		27	1:38	
29-Sep		28	1:34	
30-Sep		29	1:30	
1-Oct	New Moon	0	1:26	
2-Oct		1	1:22	
3-Oct		2	1:18	
4-Oct		3	1:14	
5-Oct		4	1:10	
6-Oct		5	1:06	
7-Oct		6	1:03	

Date	Day of Walking	Day of the Moon	Time of Calling	Notes *(Times in Ephemeris Time)*
8-Oct		7	12:59 PM	
9-Oct		8	12:55	
10-Oct	Fast Begins	9	12:51	
11-Oct		10	12:47	
12-Oct		11	12:43	
13-Oct	Black Fast	12	12:39	
14-Oct		13	12:35	
15-Oct		14	12:31	
16-Oct	Full Moon	15	12:27	
17-Oct		16	12:23	
18-Oct		17	12:19	
19-Oct		18	12:15	
20-Oct		19	12:11	
21-Oct		20	12:07	
22-Oct		21	12:03	
23-Oct		22	12:00 PM	
24-Oct		23	11:56 AM	
25-Oct		24	11:52	
26-Oct		25	11:48	
27-Oct		26	11:44	
28-Oct		27	11:40	

29-Oct		28	11:36	
30-Oct	New Moon	0	11:32	
31-Oct		1	11:28	Samhain
1-Nov		2	11:24	
2-Nov		3	11:20	
3-Nov		4	11:16	
4-Nov		5	11:12	
5-Nov		6	11:08	
6-Nov		7	11:04	
7-Nov		8	11:00	
8-Nov	Fast Begins	9	10:57	
9-Nov		10	10:53	
10-Nov		11	10:49	
11-Nov	Black Fast	12	10:45	
12-Nov		13	10:41	
13-Nov		14	10:37	
14-Nov	Full Moon	15	10:33	
15-Nov		16	10:29	
16-Nov		17	10:25	
17-Nov		18	10:21	
18-Nov		19	10:17	
19-Nov		20	10:13	
20-Nov		21	10:09	
21-Nov		22	10:05	

Date	Day of Walking	Day of the Moon	Time of Calling	Notes *(Times in Ephemeris Time)*
22-Nov		23	10:01 AM	
23-Nov		24	9:57	
24-Nov		25	9:54	
25-Nov		26	9:50	
26-Nov		27	9:46	
27-Nov		28	9:42	
28-Nov		29	9:38	
29-Nov	New Moon	0	9:34	
30-Nov		1	9:30	
1-Dec		2	9:26	
2-Dec		3	9:22	
3-Dec		4	9:18	
4-Dec		5	9:14	
5-Dec		6	9:10	
6-Dec		7	9:06	
7-Dec		8	9:02	
8-Dec	Fast Begins	9	8:58	
9-Dec		10	8:54	
10-Dec		11	8:50	
11-Dec	Black Fast	12	8:46	
12-Dec		13	8:42	

	13-Dec		14	8:38	
	14-Dec	Full Moon	15	8:34	
	15-Dec		16	8:30	
	16-Dec		17	8:26	
	17-Dec		18	8:22	
	18-Dec		19	8:18	
	19-Dec		20	8:14	
	20-Dec		21	8:10	
	21-Dec		22	8:06	Winter Solstice 10:45 AM
	22-Dec		23	8:02	
	23-Dec		24	7:58	
	24-Dec		25	7:54	
	25-Dec		26	7:51	
	26-Dec		27	7:47	
	27-Dec		28	7:43	
	28-Dec		29	7:39	
	29-Dec	New Moon	0	7:35	
	30-Dec		1	7:31	
	31-Dec		2	7:27	
2017	1-Jan		3	7:23	
	2-Jan		4	7:19	
	3-Jan		5	7:15	
	4-Jan		6	7:11	

Date	Day of Walking	Day of the Moon	Time of Calling	Notes (*Times in Ephemeris Time*)
5-Jan		7	7:07 AM	
6-Jan	Fast Begins	8	7:03	
7-Jan		9	6:59	
8-Jan		10	6:55	
9-Jan	Black Fast	11	6:51	
10-Jan		12	6:47	
11-Jan		13	6:44	
12-Jan	Full Moon	14	6:40	
13-Jan		15	6:36	
14-Jan		16	6:32	
15-Jan		17	6:28	
16-Jan		18	6:24	
17-Jan		19	6:20	
18-Jan		20	6:16	
19-Jan		21	6:12	
20-Jan		22	6:08	
21-Jan		23	6:04	
22-Jan		24	6:00	
23-Jan		25	5:56	
24-Jan		26	5:52	
25-Jan		27	5:48	

26-Jan		28	5:45	
27-Jan		29	5:41	
28-Jan	New Moon	0	5:37	
29-Jan		1	5:33	
30-Jan		2	5:29	
31-Jan		3	5:25	Imbolc
1-Feb		4	5:21	
2-Feb		5	5:17	
3-Feb		6	5:13	
4-Feb		7	5:09	
5-Feb	Fast Begins	8	5:05	
6-Feb		9	5:01	
7-Feb		10	4:57	
8-Feb	Black Fast	11	4:53	
9-Feb		12	4:49	
10-Feb		13	4:45	
11-Feb	Full Moon	14	4:42	Appulse Lunar Eclipse 12:45 AM
12-Feb		15	4:38	
13-Feb		16	4:34	
14-Feb		17	4:30	
15-Feb		18	4:26	
16-Feb		19	4:22	
17-Feb		20	4:18	
18-Feb		21	4:14	

Date	Day of Walking	Day of the Moon	Time of Calling	Notes (*Times in Ephemeris Time*)
19-Feb		22	4:10 AM	
20-Feb		23	4:06	
21-Feb		24	4:02	
22-Feb		25	3:58	
23-Feb		26	3:54	
24-Feb		27	3:50	
25-Feb		28	3:46	
26-Feb	New Moon	0	3:42	Annular Solar Eclipse 2:54 PM
27-Feb		1	3:38	
28-Feb		2	3:34	
1-Mar		3	3:30	
2-Mar		4	3:26	
3-Mar		5	3:22	
4-Mar		6	3:18	
5-Mar		7	3:14	
6-Mar	Fast Begins	8	3:10	
7-Mar		9	3:06	
8-Mar		10	3:02	
9-Mar	Black Fast	11	2:58	
10-Mar		12	2:54	
11-Mar		13	2:50	

12-Mar	Full Moon	14	2:46	
13-Mar		15	2:42	
14-Mar		16	2:38	
15-Mar		17	2:34	
16-Mar		18	2:30	
17-Mar		19	2:26	
18-Mar		20	2:22	
19-Mar		21	2:18	
20-Mar		22	2:14	Vernal Equinox 10:30 AM
21-Mar		23	2:10	
22-Mar		24	2:06	
23-Mar		25	2:02	
24-Mar		26	1:58	
25-Mar		27	1:54	
26-Mar		28	1:50	
27-Mar		29	1:46	
28-Mar	New Moon	0	1:42	
29-Mar		1	1:38	
30-Mar		2	1:34	
31-Mar		3	1:30	
1-Apr		4	1:26	
2-Apr		5	1:22	
3-Apr		6	1:18	
4-Apr		7	1:14	

Date	Day of Walking	Day of the Moon	Time of Calling	Notes (*Times in Ephemeris Time*)
5-Apr	Fast Begins	8	1:10 AM	
6-Apr		9	1:06	
7-Apr		10	1:02	
8-Apr	Black Fast	11	12:58	
9-Apr		12	12:54	
10-Apr		13	12:50	
11-Apr	Full Moon	14	12:46	
12-Apr		15	12:42	
13-Apr		16	12:38	
14-Apr		17	12:34	
15-Apr		18	12:30	
16-Apr		19	12:25	
17-Apr		20	12:21	
18-Apr		21	12:17	
19-Apr		22	12:13	
20-Apr		23	12:09	
21-Apr		24	12:05	
22-Apr		25	12:01 AM	
23-Apr		26	11:57 PM	
24-Apr		27	11:53	
25-Apr		28	11:49	

26-Apr	New Moon	0	11:45	
27-Apr		1	11:41	
28-Apr		2	11:37	
29-Apr		3	11:33	
30-Apr		4	11:30	Beltane
1-May		5	11:26	
2-May		6	11:22	
3-May		7	11:18	
4-May	Fast Begins	8	11:14	
5-May		9	11:10	
6-May		10	11:06	
7-May	Black Fast	11	11:02	
8-May		12	10:58	
9-May		13	10:54	
10-May	Full Moon	14	10:50	
11-May		15	10:46	
12-May		16	10:42	
13-May		17	10:38	
14-May		18	10:34	
15-May		19	10:31	
16-May		20	10:27	
17-May		21	10:23	
18-May		22	10:19	
19-May		23	10:15	

Date	Day of Walking	Day of the Moon	Time of Calling	Notes (*Times in Ephemeris Time*)
20-May		24	10:11 PM	
21-May		25	10:07	
22-May		26	10:03	
23-May		27	9:59	
24-May		28	9:55	
25-May	New Moon	0	9:51	
26-May		1	9:47	
27-May		2	9:43	
28-May		3	9:39	
29-May		4	9:35	
30-May		5	9:31	
31-May		6	9:28	
1-Jun		7	9:24	
2-Jun		8	9:20	
3-Jun	Fast Begins	9	9:16	
4-Jun		10	9:12	
5-Jun		11	9:08	
6-Jun	Black Fast	12	9:04	
7-Jun		13	9:00	
8-Jun		14	8:56	
9-Jun	Full Moon	15	8:52	

10-Jun		16	8:48	
11-Jun		17	8:44	
12-Jun		18	8:40	
13-Jun		19	8:36	
14-Jun		20	8:32	
15-Jun		21	8:28	
16-Jun		22	8:25	
17-Jun		23	8:21	
18-Jun		24	8:17	
19-Jun		25	8:13	
20-Jun		26	8:09	
21-Jun		27	8:05	Summer Solstice 4:25 AM
22-Jun		28	8:01	
23-Jun		29	7:57	
24-Jun	New Moon	0	7:53	
25-Jun		1	7:49	
26-Jun		2	7:45	
27-Jun		3	7:41	
28-Jun		4	7:37	
29-Jun		5	7:33	
30-Jun		6	7:29	
1-Jul		7	7:25	
2-Jul		8	7:22	
3-Jul	Fast Begins	9	7:18	

Date	Day of Walking	Day of the Moon	Time of Calling	Notes *(Times in Ephemeris Time)*
4-Jul		10	7:14 PM	
5-Jul		11	7:10	
6-Jul	Black Fast	12	7:06	
7-Jul		13	7:02	
8-Jul		14	6:58	
9-Jul	Full Moon	15	6:54	
10-Jul		16	6:50	
11-Jul		17	6:46	
12-Jul		18	6:42	
13-Jul		19	6:38	
14-Jul		20	6:34	
15-Jul		21	6:30	
16-Jul		22	6:26	
17-Jul		23	6:22	
18-Jul		24	6:19	
19-Jul		25	6:15	
20-Jul		26	6:11	
21-Jul		27	6:07	
22-Jul		28	6:03	
23-Jul	New Moon	0	5:59	
24-Jul		1	5:55	

25-Jul		2	5:51	
26-Jul		3	5:47	
27-Jul		4	5:43	
28-Jul		5	5:39	
29-Jul		6	5:35	
30-Jul		7	5:31	
31-Jul		8	5:27	
1-Aug	Fast Begins	9	5:23	Lammas
2-Aug		10	5:19	
3-Aug		11	5:18	
4-Aug	Black Fast	12	5:14	
5-Aug		13	5:10	
6-Aug		14	5:06	
7-Aug	Full Moon	15	5:02	Partial Lunar Eclipse 6:22 PM
8-Aug		16	4:58	
9-Aug		17	4:54	
10-Aug		18	4:50	
11-Aug		19	4:46	
12-Aug		20	4:42	
13-Aug		21	4:38	
14-Aug		22	4:34	
15-Aug		23	4:30	
16-Aug		24	4:26	
17-Aug		25	4:22	

Date	Day of Walking	Day of the Moon	Time of Calling	Notes *(Times in Ephemeris Time)*
18-Aug		26	4:18 PM	
19-Aug		27	4:16	
20-Aug		28	4:12	
21-Aug	New Moon	0	4:08	Total Solar Eclipse 6:26 PM
22-Aug		1	4:04	
23-Aug		2	4:00	
24-Aug		3	3:56	
25-Aug		4	3:52	
26-Aug		5	3:48	
27-Aug		6	3:44	
28-Aug		7	3:40	
29-Aug		8	3:36	
30-Aug		9	3:32	
31-Aug	Fast Begins	10	3:28	
1-Sep		11	3:24	
2-Sep		12	3:20	
3-Sep	Black Fast	13	3:16	
4-Sep		14	3:12	
5-Sep		15	3:09	
6-Sep	Full Moon	16	3:05	
7-Sep		17	3:01	

8-Sep		18	2:57	
9-Sep		19	2:53	
10-Sep		20	2:49	
11-Sep		21	2:45	
12-Sep		22	2:41	
13-Sep		23	2:37	
14-Sep		24	2:33	
15-Sep		25	2:29	
16-Sep		26	2:25	
17-Sep		27	2:21	
18-Sep		28	2:17	
19-Sep		29	2:13	
20-Sep	New Moon	0	2:09	
21-Sep		1	2:06	
22-Sep		2	2:02	Autumnal Equinox 8:03 PM
23-Sep		3	1:58	
24-Sep		4	1:54	
25-Sep		5	1:50	
26-Sep		6	1:46	
27-Sep		7	1:42	
28-Sep		8	1:38	
29-Sep	Fast Begins	9	1:34	
30-Sep		10	1:30	
1-Oct		11	1:26	

Date	Day of Walking	Day of the Moon	Time of Calling	Notes (*Times in Ephemeris Time*)
2-Oct	Black Fast	12	1:22 PM	
3-Oct		13	1:18	
4-Oct		14	1:14	
5-Oct	Full Moon	15	1:10	
6-Oct		16	1:06	
7-Oct		17	1:03	
8-Oct		18	12:59	
9-Oct		19	12:55	
10-Oct		20	12:51	
11-Oct		21	12:47	
12-Oct		22	12:43	
13-Oct		23	12:39	
14-Oct		24	12:35	
15-Oct		25	12:31	
16-Oct		26	12:27	
17-Oct		27	12:23	
18-Oct		28	12:19	
19-Oct	New Moon	0	12:15	
20-Oct		1	12:11	
21-Oct		2	12:07	
22-Oct		3	12:03	

23-Oct		4	12:00 PM	
24-Oct		5	11:56 AM	
25-Oct		6	11:52	
26-Oct		7	11:48	
27-Oct		8	11:44	
28-Oct		9	11:40	
29-Oct	Fast Begins	10	11:36	
30-Oct		11	11:32	
31-Oct		12	11:28	Samhain
1-Nov	Black Fast	13	11:24	
2-Nov		14	11:20	
3-Nov		15	11:16	
4-Nov	Full Moon	16	11:12	
5-Nov		17	11:08	
6-Nov		18	11:04	
7-Nov		19	11:00	
8-Nov		20	10:57	
9-Nov		21	10:53	
10-Nov		22	10:49	
11-Nov		23	10:45	
12-Nov		24	10:41	
13-Nov		25	10:37	
14-Nov		26	10:33	
15-Nov		27	10:29	

Date	Day of Walking	Day of the Moon	Time of Calling	Notes *(Times in Ephemeris Time)*
16-Nov		28	10:25 AM	
17-Nov		29	10:21	
18-Nov	New Moon	0	10:17	
19-Nov		1	10:13	
20-Nov		2	10:09	
21-Nov		3	10:05	
22-Nov		4	10:01	
23-Nov		5	9:57	
24-Nov		6	9:54	
25-Nov		7	9:50	
26-Nov		8	9:46	
27-Nov	Fast Begins	9	9:42	
28-Nov		10	9:38	
29-Nov		11	9:34	
30-Nov	Black Fast	12	9:30	
1-Dec		13	9:26	
2-Dec		14	9:22	
3-Dec	Full Moon	15	9:18	
4-Dec		16	9:14	
5-Dec		17	9:10	
6-Dec		18	9:06	

7-Dec		19	9:02	
8-Dec		20	8:58	
9-Dec		21	8:54	
10-Dec		22	8:50	
11-Dec		23	8:46	
12-Dec		24	8:42	
13-Dec		25	8:38	
14-Dec		26	8:34	
15-Dec		27	8:30	
16-Dec		28	8:26	
17-Dec		29	8:22	
18-Dec	New Moon	0	8:18	
19-Dec		1	8:14	
20-Dec		2	8:10	
21-Dec		3	8:06	Winter Solstice 4:29 PM
22-Dec		4	8:02	
23-Dec		5	7:58	
24-Dec		6	7:54	
25-Dec		7	7:51	
26-Dec		8	7:47	
27-Dec	Fast Begins	9	7:43	
28-Dec		10	7:39	
29-Dec		11	7:35	
30-Dec	Black Fast	12	7:31	

Date		Day of Walking	Day of the Moon	Time of Calling	Notes *(Times in Ephemeris Time)*
	31-Dec		13	7:27 AM	
2018	1-Jan		14	7:23	
	2-Jan	Full Moon	15	7:19	
	3-Jan		16	7:15	
	4-Jan		17	7:11	
	5-Jan		18	7:07	
	6-Jan		19	7:03	
	7-Jan		20	6:59	
	8-Jan		21	6:55	
	9-Jan		22	6:51	
	10-Jan		23	6:47	
	11-Jan		24	6:44	
	12-Jan		25	6:40	
	13-Jan		26	6:36	
	14-Jan		27	6:32	
	15-Jan		28	6:28	
	16-Jan		29	6:24	
	17-Jan	New Moon	0	6:20	
	18-Jan		1	6:16	
	19-Jan		2	6:12	

20-Jan		3	6:08	
21-Jan		4	6:04	
22-Jan		5	6:00	
23-Jan		6	5:56	
24-Jan		7	5:52	
25-Jan	Fast Begins	8	5:48	
26-Jan		9	5:45	
27-Jan		10	5:41	
28-Jan	Black Fast	11	5:37	
29-Jan		12	5:33	
30-Jan		13	5:29	
31-Jan	Full Moon	14	5:25	Total Lunar Eclipse 1:31 PM / Imbolc
1-Feb		15	5:21	
2-Feb		16	5:17	
3-Feb		17	5:13	
4-Feb		18	5:09	
5-Feb		19	5:05	
6-Feb		20	5:01	
7-Feb		21	4:57	
8-Feb		22	4:53	
9-Feb		23	4:49	
10-Feb		24	4:45	
11-Feb		25	4:42	

Date	Day of Walking	Day of the Moon	Time of Calling	Notes *(Times in Ephemeris Time)*
12-Feb		26	4:38 AM	
13-Feb		27	4:34	
14-Feb		28	4:30	
15-Feb	New Moon	0	4:26	Partial Solar Eclipse 8:52 PM
16-Feb		1	4:22	
17-Feb		2	4:18	
18-Feb		3	4:14	
19-Feb		4	4:10	
20-Feb		5	4:06	
21-Feb		6	4:02	
22-Feb		7	3:58	
23-Feb		8	3:54	
24-Feb	Fast Begins	9	3:50	
25-Feb		10	3:46	
26-Feb		11	3:42	
27-Feb	Black Fast	12	3:38	
28-Feb		13	3:34	
1-Mar		14	3:30	
2-Mar	Full Moon	15	3:26	
3-Mar		16	3:22	
4-Mar		17	3:18	

5-Mar		18	3:14	
6-Mar		19	3:10	
7-Mar		20	3:06	
8-Mar		21	3:02	
9-Mar		22	2:58	
10-Mar		23	2:54	
11-Mar		24	2:50	
12-Mar		25	2:46	
13-Mar		26	2:42	
14-Mar		27	2:38	
15-Mar		28	2:34	
16-Mar		29	2:30	
17-Mar	New Moon	0	2:26	
18-Mar		1	2:22	
19-Mar		2	2:18	
20-Mar		3	2:14	Vernal Equinox 4:17 PM
21-Mar		4	2:10	
22-Mar		5	2:06	
23-Mar		6	2:02	
24-Mar		7	1:58	
25-Mar	Fast Begins	8	1:54	
26-Mar		9	1:50	
27-Mar		10	1:46	
28-Mar	Black Fast	11	1:42	

Date	Day of Walking	Day of the Moon	Time of Calling	Notes *(Times in Ephemeris Time)*
29-Mar		12	1:38 AM	
30-Mar		13	1:34	
31-Mar	Full Moon	14	1:30	
1-Apr		15	1:26	
2-Apr		16	1:22	
3-Apr		17	1:18	
4-Apr		18	1:14	
5-Apr		19	1:10	
6-Apr		20	1:06	
7-Apr		21	1:02	
8-Apr		22	12:58	
9-Apr		23	12:54	
10-Apr		24	12:50	
11-Apr		25	12:46	
12-Apr		26	12:42	
13-Apr		27	12:38	
14-Apr		28	12:34	
15-Apr		29	12:30	
16-Apr	New Moon	0	12:25	
17-Apr		1	12:21	
18-Apr		2	12:17	

19-Apr		3	12:13	
20-Apr		4	12:09	
21-Apr		5	12:05	
22-Apr		6	12:01 AM	
23-Apr		7	11:57 PM	
24-Apr	Fast Begins	8	11:53	
25-Apr		9	11:49	
26-Apr		10	11:45	
27-Apr	Black Fast	11	11:41	
28-Apr		12	11:37	
29-Apr		13	11:33	
30-Apr	Full Moon	14	11:30	Beltane
1-May		15	11:26	
2-May		16	11:22	
3-May		17	11:18	
4-May		18	11:14	
5-May		19	11:10	
6-May		20	11:06	
7-May		21	11:02	
8-May		22	10:58	
9-May		23	10:54	
10-May		24	10:50	
11-May		25	10:46	
12-May		26	10:42	

Date	Day of Walking	Day of the Moon	Time of Calling	Notes (*Times in Ephemeris Time*)
13-May		27	10:38 PM	
14-May		28	10:34	
15-May	New Moon	0	10:31	
16-May		1	10:27	
17-May		2	10:23	
18-May		3	10:19	
19-May		4	10:15	
20-May		5	10:11	
21-May		6	10:07	
22-May		7	10:03	
23-May	Fast Begins	8	9:59	
24-May		9	9:55	
25-May		10	9:51	
26-May	Black Fast	11	9:47	
27-May		12	9:43	
28-May		13	9:39	
29-May	Full Moon	14	9:35	
30-May		15	9:31	
31-May		16	9:28	
1-Jun		17	9:24	
2-Jun		18	9:20	

3-Jun		19	9:16	
4-Jun		20	9:12	
5-Jun		21	9:08	
6-Jun		22	9:04	
7-Jun		23	9:00	
8-Jun		24	8:56	
9-Jun		25	8:52	
10-Jun		26	8:48	
11-Jun		27	8:44	
12-Jun		28	8:40	
13-Jun	New Moon	0	8:36	
14-Jun		1	8:32	
15-Jun		2	8:28	
16-Jun		3	8:25	
17-Jun		4	8:21	
18-Jun		5	8:17	
19-Jun		6	8:13	
20-Jun		7	8:09	
21-Jun		8	8:05	Summer Solstice 10:08 AM
22-Jun	Fast Begins	9	8:01	
23-Jun		10	7:57	
24-Jun		11	7:53	
25-Jun	Black Fast	12	7:49	
26-Jun		13	7:45	

Date	Day of Walking	Day of the Moon	Time of Calling	Notes (*Times in Ephemeris Time*)
27-Jun		14	7:41 PM	
28-Jun	Full Moon	15	7:37	
29-Jun		16	7:33	
30-Jun		17	7:29	
1-Jul		18	7:25	
2-Jul		19	7:22	
3-Jul		20	7:18	
4-Jul		21	7:14	
5-Jul		22	7:10	
6-Jul		23	7:06	
7-Jul		24	7:02	
8-Jul		25	6:58	
9-Jul		26	6:54	
10-Jul		27	6:50	
11-Jul		28	6:46	
12-Jul		29	6:42	
13-Jul	New Moon	0	6:38	Partial Solar Eclipse 9:47 AM
14-Jul		1	6:34	
15-Jul		2	6:30	
16-Jul		3	6:26	
17-Jul		4	6:22	

18-Jul		5	6:19	
19-Jul		6	6:15	
20-Jul		7	6:11	
21-Jul	Fast Begins	8	6:07	
22-Jul		9	6:03	
23-Jul		10	5:59	
24-Jul	Black Fast	11	5:55	
25-Jul		12	5:51	
26-Jul		13	5:47	
27-Jul	Full Moon	14	5:43	Total Lunar Eclipse 8:23 PM
28-Jul		15	5:39	
29-Jul		16	5:35	
30-Jul		17	5:31	
31-Jul		18	5:27	
1-Aug		19	5:23	Beltane
2-Aug		20	5:19	
3-Aug		21	5:18	
4-Aug		22	5:14	
5-Aug		23	5:10	
6-Aug		24	5:06	
7-Aug		25	5:02	
8-Aug		26	4:58	
9-Aug		27	4:54	
10-Aug		28	4:50	

Date	Day of Walking	Day of the Moon	Time of Calling	Notes *(Times in Ephemeris Time)*
11-Aug	New Moon	0	4:46 PM	Partial Solar Eclipse 9:47 AM
12-Aug		1	4:42	
13-Aug		2	4:38	
14-Aug		3	4:34	
15-Aug		4	4:30	
16-Aug		5	4:26	
17-Aug		6	4:22	
18-Aug		7	4:18	
19-Aug		8	4:16	
20-Aug	Fast Begins	9	4:12	
21-Aug		10	4:08	
22-Aug		11	4:04	
23-Aug	Black Fast	12	4:00	
24-Aug		13	3:56	
25-Aug		14	3:52	
26-Aug	Full Moon	15	3:48	
27-Aug		16	3:44	
28-Aug		17	3:40	
29-Aug		18	3:36	
30-Aug		19	3:32	
31-Aug		20	3:28	

1-Sep		21	3:24	
2-Sep		22	3:20	
3-Sep		23	3:16	
4-Sep		24	3:12	
5-Sep		25	3:09	
6-Sep		26	3:05	
7-Sep		27	3:01	
8-Sep		28	2:57	
9-Sep	New Moon	0	2:53	
10-Sep		1	2:49	
11-Sep		2	2:45	
12-Sep		3	2:41	
13-Sep		4	2:37	
14-Sep		5	2:33	
15-Sep		6	2:29	
16-Sep		7	2:25	
17-Sep		8	2:21	
18-Sep		9	2:17	
19-Sep	Fast Begins	10	2:13	
20-Sep		11	2:09	
21-Sep		12	2:06	
22-Sep	Black Fast	13	2:02	
23-Sep		14	1:58	Autumnal Equinox 1:55 AM
24-Sep		15	1:54	

Date	Day of Walking	Day of the Moon	Time of Calling	Notes *(Times in Ephemeris Time)*
25-Sep	Full Moon	16	1:50 PM	
26-Sep		17	1:46	
27-Sep		18	1:42	
28-Sep		19	1:38	
29-Sep		20	1:34	
30-Sep		21	1:30	
1-Oct		22	1:26	
2-Oct		23	1:22	
3-Oct		24	1:18	
4-Oct		25	1:14	
5-Oct		26	1:10	
6-Oct		27	1:06	
7-Oct		28	1:03	
8-Oct		29	12:59	
9-Oct	New Moon	0	12:55	
10-Oct		1	12:51	
11-Oct		2	12:47	
12-Oct		3	12:43	
13-Oct		4	12:39	
14-Oct		5	12:35	
15-Oct		6	12:31	

16-Oct		7	12:27	
17-Oct		8	12:23	
18-Oct	Fast Begins	9	12:19	
19-Oct		10	12:15	
20-Oct		11	12:11	
21-Oct	Black Fast	12	12:07	
22-Oct		13	12:03	
23-Oct		14	12:00 PM	
24-Oct	Full Moon	15	11:56 AM	
25-Oct		16	11:52	
26-Oct		17	11:48	
27-Oct		18	11:44	
28-Oct		19	11:40	
29-Oct		20	11:36	
30-Oct		21	11:32	
31-Oct		22	11:28	Samhain
1-Nov		23	11:24	
2-Nov		24	11:20	
3-Nov		25	11:16	
4-Nov		26	11:12	
5-Nov		27	11:08	
6-Nov		28	11:04	
7-Nov	New Moon	0	11:00	
8-Nov		1	10:57	

Date	Day of Walking	Day of the Moon	Time of Calling	Notes (*Times in Ephemeris Time*)
9-Nov		2	10:53 AM	
10-Nov		3	10:49	
11-Nov		4	10:45	
12-Nov		5	10:41	
13-Nov		6	10:37	
14-Nov		7	10:33	
15-Nov		8	10:29	
16-Nov		9	10:25	
17-Nov	Fast Begins	10	10:21	
18-Nov		11	10:17	
19-Nov		12	10:13	
20-Nov	Black Fast	13	10:09	
21-Nov		14	10:05	
22-Nov		15	10:01	
23-Nov	Full Moon	16	9:57	
24-Nov		17	9:54	
25-Nov		18	9:50	
26-Nov		19	9:46	
27-Nov		20	9:42	
28-Nov		21	9:38	
29-Nov		22	9:34	

30-Nov		23	9:30	
1-Dec		24	9:26	
2-Dec		25	9:22	
3-Dec		26	9:18	
4-Dec		27	9:14	
5-Dec		28	9:10	
6-Dec		29	9:06	
7-Dec	New Moon	0	9:02	
8-Dec		1	8:58	
9-Dec		2	8:54	
10-Dec		3	8:50	
11-Dec		4	8:46	
12-Dec		5	8:42	
13-Dec		6	8:38	
14-Dec		7	8:34	
15-Dec		8	8:30	
16-Dec	Fast Begins	9	8:26	
17-Dec		10	8:22	
18-Dec		11	8:18	
19-Dec	Black Fast	12	8:14	
20-Dec		13	8:10	
21-Dec		14	8:06	Winter Solstice 10:24 PM
22-Dec	Full Moon	15	8:02	
23-Dec		16	7:58	

Date	Day of Walking	Day of the Moon	Time of Calling	Notes *(Times in Ephemeris Time)*
24-Dec		17	7:54 AM	
25-Dec		18	7:51	
26-Dec		19	7:47	
27-Dec		20	7:43	
28-Dec		21	7:39	
29-Dec		22	7:35	
30-Dec		23	7:31	
31-Dec		24	7:27	

Table of Chinese Terms

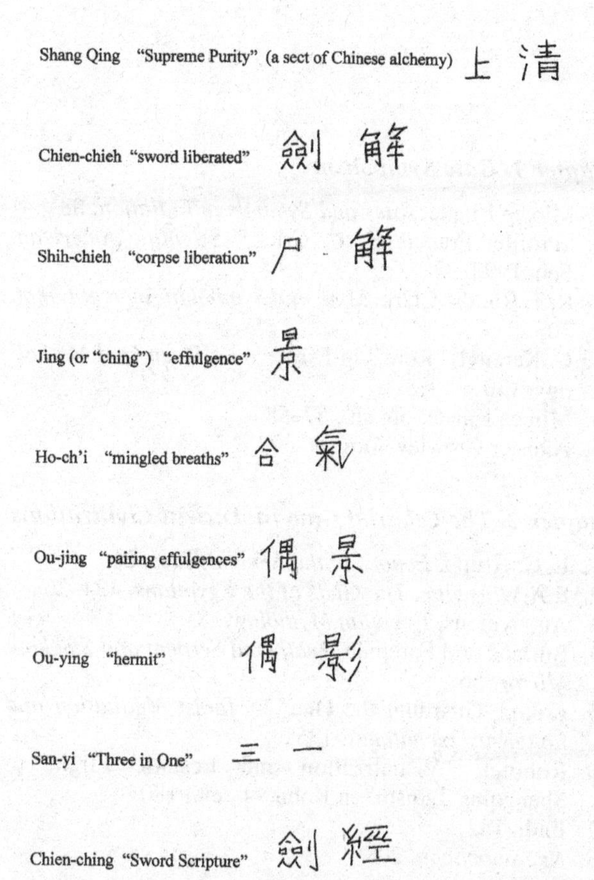

Shang Qing "Supreme Purity" (a sect of Chinese alchemy) 上清

Chien-chieh "sword liberated" 劍解

Shih-chieh "corpse liberation" 尸解

Jing (or "ching") "effulgence" 景

Ho-ch'i "mingled breaths" 合氣

Ou-jing "pairing effulgences" 偶景

Ou-ying "hermit" 偶影

San-yi "Three in One" 三一

Chien-ching "Sword Scripture" 劍經

Notes

Chapter 1. Gate Symbolism

1. Mircea Eliade, *Rites and Symbols of Initiation*, 88
2. "Profile: Francis H. C. Crick," *Scientific American*, Feb. 1992, 33
3. R. T. Rundle Clark, *Myth and Symbol in Ancient Egypt*, 205
4. C. Kerenyi, "Kore," in *Essays on a Science of Mythology*, 130
5. Mircea Eliade, op. cit., 57–58
6. Aleister Crowley, *Magick*, 328

Chapter 2. The Celestial Gate in Ancient Civilizations

1. E. C. Krupp, *Echoes of the Ancient Skies*, 26
2. E.A.W. Budge, *The Gods of the Egyptians*, 424–26
3. Anonymous, *Egyptian Mythology*, 85
4. Burland and Foreman, *Feathered Serpent and Smoking Mirror*, 56
5. Kohn, "Guarding the One" in *Taoist Meditation and Longevity Techniques*, 135
6. Robinet, "Visualization and Ecstatic Flight in Shangqing Taoism" in Kohn, op. cit., 180
7. Ibid., 182
8. *Necronomicon*, 208

9. Ibid., 210
10. Clark, op. cit., 52 and 141
11. Daniélou, *The Gods of India: Hindu Polytheism*, 316–17
12. Ibid., 319
13. Ibid., 317
14. Ibid., 308
15. Clark, op. cit., 167
16. Evans-Wentz, *The Tibetan Book of the Dead*, 117
17. Clark, op. cit., 151
18. Crowley, *Goetia*, 37
19. E.A.W. Budge, *The Book of Opening The Mouth*, 67
20. E.A.W. Budge, *From Fetish to God in Ancient Egypt*, 34
21. Budge, *The Book of Opening The Mouth*, 67
22. Ibid., 68
23. Ibid., 69. Author Lucie Lamy in *Egyptian Mysteries* states several times that the adze or pair of adzes were made of magnetite, thus reinforcing my thesis, but does not give her references for this crucial piece of information. If the adze was indeed magnetic, it would indicate that the ancient Egyptians were aware of magnetism and used magnetic instruments in much the same way as the Chinese diviners: as instruments sacred to the Dipper and to the god(s) of the Dipper, useful for orienting temples and tombs.
24. Ibid., 69
25. Ibid., 70
26. Ibid., 70
27. Taylor, "The Gundestrup Cauldron," *Scientific American*, March 1992, 84–90
28. Budge, *The Book of Opening The Mouth*, 70–71
29. Ibid., 72–73
30. Ibid., 71
31. Clark, op. cit., 241
32. Ibid., 160
33. see R. D. Laing, *The Politics of Experience*
34. see Parkinson, *Parkinson's Law*

Chapter 3. The Great Bear: Key to the Gates

1. see Temple, *The Sirius Mystery*
2. von Franz, "The Process of Individuation," in *Man And His Symbols*, Jung, ed., 171
3. Walters, *Chinese Astrology*, 160
4. *The American Heritage Dictionary of Indo-European Roots*, Watkins, ed., 55
5. MacCana, *Celtic Mythology*, 48
6. Walters, op. cit., 93

Chapter 4. The Great Bear in Shangqing Daoism

1. Strickmann, "On The Alchemy of T'ao Hung-Ching" in *Facets of Taoism*, 189
2. Temple, op. cit., 185
3. Kohn, "Guarding the One" in Kohn, op. cit., 142
4. Ibid., 139
5. Temple, op. cit., 20, 22, 40 (compare with the description of the "circumambulation" in *Magick*, 347)
6. Strickmann, op. cit., 173
7. Ibid., 173
8. see Blofeld, *The Wheel of Life* and *Taoist Mysteries and Magic*

Chapter 5. The Formation of the Astral Body

1. Strickmann, op. cit., 174–75
2. Crowley, *Magick*, XII
3. *Necronomicon*, 47

Chapter 6. The Primordial Conflict

1. see Paul Davies, *The Mind of God*
2. Neely, *A Primer For Star-Gazers*, 102
3. For references to this common identification of serpents with wisdom and "wise men," see Temple, op.

cit., and the works of authors such as G.R.S. Mead and others on the Ophites and Middle Eastern Gnostic cults, as well as the Old Testament.

Tables of the Bear

1. Kramer, *The Sumerians*, 91

Bibliography

Works Cited in Notes

Anon., *Egyptian Mythology*, Tudor Publishing Co., NY, 1965

Anon., *Necronomicon*, Schlangekraft Publishing Co., NY, 1977

— *Necronomicon*, Avon Books, NY, 1980

Blofeld, John, *Taoist Mysteries and Magic*, Shambhala, Boulder, 1982

Budge, E.A. Wallis, *From Fetish To God in Ancient Egypt*, Oxford University Press, London, 1934

— *The Book of Opening The Mouth*, Kegan Paul, Trench, Trübner & Co., Ltd., London, 1909

— *The Gods of the Egyptians*, The Open Court Publishing Co., Chicago, 1904

Burland, C.A. and Forman, Werner, *Feathered Serpent and Smoking Mirror*, G.P. Putnam's Sons, NY, 1975

Clark, R.T. Rundle, *Myth and Symbol in Ancient Egypt*, Thames & Hudson, NY, 1991

Crowley, Aleister, *Magick In Theory and Practice*, Magickal Childe Publishing Co., NY, 1990

— *Goetia*, Magickal Childe Publishing Co., NY, 1989

Danielou, Alain, *The Gods of India: Hindu Polytheism*, Inner Traditions International, Ltd., NY, 1985